THE

SLIDE - RULE,

AND

HOW TO USE IT:

CONTAINING

FULL, EASY, AND SIMPLE INSTRUCTIONS TO PERFORM

ALL BUSINESS CALCULATIONS

WITH

UNEXAMPLED RAPIDITY AND ACCURACY.

BY

CHARLES HOARE, C.E.,

AUTHOR OF "MENSURATION MADE EASY," ETC. ETC.

With a Slide-Rule in Tuck of Cover.

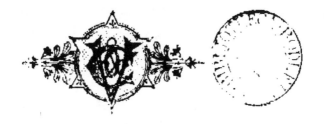

LONDON:

VIRTUE & CO., 26, IVY LANE,

PATERNOSTER ROW.

1868.

PREFACE.

To those who have once acquired a knowledge of the capabilities of the Slide Rule, it is ever a matter of surprise that an instrument combining such unexampled rapidity, ease, and accuracy in performing all ordinary business calculations, should be so little known. By its assistance the drudgery of computation is avoided, and the time and trouble expended on mere arithmetical workings proved to be a waste of effort; in fact, its aid mentally may safely be compared with the advantages derived from mechanical appliances in ameliorating the wear and tear of manual labour.

The intellect remains unfettered by tedious processes, for the statement of each question, the operation and the result, are simultaneous and apparent in their connection. The laws that govern its operations are few and simple, and easily understood; and the curiosity of the uninitiated may be stimulated by learning, that on an instrument as portable as a pocket-book we have the whole gamut of numbers; and that whether as a means for self-instruction or advancement, for unsurpassed utility in business, or for profitable amusement, its study is well rewarded in its capabilities for varied application. Scientific men estimate its value, the man of business would soon appreciate its assistance, and it will be well for the practical mechanic when he learns how to employ it intelligently, instead of carrying it in his pocket, yet unable to avail himself of its extraordinary powers.

The disuse of the Slide Rule in ordinary calculations, in face of its proved capabilities, suggested the idea that either its construction, or the method of teaching, or perhaps both, might be capable of amendment. The adept may smile at the proposal to modify an instrument already simple enough to him, but there is evidence that, to make it available to many, it must first be made easy to all; for, generally speaking, its use has to be acquired by self-teaching, and if the professed instructions be not clear enough to pilot the beginner through the *seeming* difficulties of a new study, they are useless.

Sufficient introductory matter, and ample explanation, are needed to familiarise the student with the subject and its advantages. Treatises have been printed by the score, but Bevan, Woolgar, and other scientific writers, are scarcely before the public; while some are above, and many below general comprehension. To be popular, such matter must be plain. Abler pens might have invested the subject with greater interest; my aim has been simplicity of method and arrangement. Through the liberality of the present publishers of Weale's useful series, I have been unrestricted in space and detail; and believing that earnestness and accuracy may be accepted in lieu of higher pretensions, I trust that the large amount of information embodied in the work will prove acceptable and useful.

CHARLES HOARE.

SOUTHAMPTON,
   November, 1867.

CONTENTS.

THE SLIDE RULE.

INTRODUCTORY AND EXPLANATORY REMARKS.

THE combinations of the Slide Rule, like the elementary processes in Arithmetic, are few and simple, but their application is almost unlimited. Its action being mechanical, the working can happily be illustrated *without written rules*, in lieu of which, a copy of the position of the lines and figures, in fact, a diagram of the statement for each class of operations, is given, and the directions fully detailed in No. 6 of these Notes. The following memoranda, necessarily ample, are descriptive merely, enforcing no tax upon the memory after the explanations of the lines, numbers, and divisions on the Rule are clearly comprehended. No pains have been spared to render these preliminary instructions as concise and clear as possible. The successful practice of all that follows depends upon their being thoroughly *understood;* in such hands the Slide Rule is an intelligible and powerful instrument ;—in others a mere tool.

1st. All numbers and divisions are to be read decimally, for all the spaces are, or are supposed to be, divided and subdivided into tens and tenths ; the visible marks may describe fifths, or halves—these are still equal parts of ten. Where the spaces do not admit of subdivision, the proportions of $\frac{1}{4}$, $\frac{1}{2}$, $\frac{3}{4}$ must be estimated ; and when the eye grows accustomed to the scale, with a little practice, tenths of a division may be judged with great accuracy.

2nd. The figures on the Rule are engraved simply as 1, 2, 3, &c. ; but these numbers are arbitrary, and *any* required value may be assigned ; thus, a 2 may be called 2 or 20, or 200 ; if it is borne in mind that the whole line is affected during that operation.

Ex.: A 2 being called 20 the 3 is 30, and so on throughout the Scale on that line, but *different lines* may bear different values if the proportions are maintained, a simple case will illustrate this point.

Ex.: If 2 bushels cost 20s., what will 6½ bushels and also 65 bushels cost?

The statement on the Rule would be—

A Set 2 bushels   6,5 bushels   and 65 bushels
B To 20 shillings   cost 65s.   cost 650 shillings

Here we have given the 2 *on the Slide* a tenfold value, viz.: 20s., which extends to the 65 and the 650.

3rd. The ORDINARY Slide Rule consists of four lines, viz.:—

A
B } Being on the Slide, the edge of B working with A.
C }   „   „   „   C   „   D.
D

This arrangement, though compact enough, is puzzling to the beginner, giving an unnecessary appearance of intricacy to a very simple instrument.

In the form of rule recommended and adopted by the Author, the pairs of lines

$\frac{A}{B}$ and $\frac{C}{D}$ } *are separated,* each pair having its own office, instead of a combined and complex operation. The following explanation of the uses and relation of these lines and slides is greatly assisted by an inspection of the Rule at the time.

$\frac{A}{B}$ Lines ; will be found exactly alike in numbers and divisions, and are, therefore, in direct simple proportion, and *all* such questions may be solved by them. When closed they stand thus :—

A 1   2   3   4 &c., &c.
B 1   2   3   4 &c., &c.

but if 1 in the Slide is projected to 2 on A, it will be seen that the ratio of ½ runs throughout.

A 1   2   4   6 &c.
B   1   2   3 &c.

and so of any proportion we choose to make between $\frac{A}{B}$ lines.

$\frac{C}{D}$ Lines are relatively different, the effect being that all numbers and divisions on C are the square or self-multiple of the numbers on D, and consequently the numbers and divisions on D are the *square roots* of those on C, thus—

C 1   4   9   16 &c., &c.
D 1   2   3   4

Questions of Area or Solidity—proportions that are as to the square or square root of numbers—are solved by them.

4th. The reading or valuing of the figures and divisions on the several lines requires a little practice; the principle is simple enough. It has been shown that a numerical value being assigned to any figure, the rest of the line is determined by it, thus,

if the left	the centre	the right
is 1	is 10	is 1000
	or	
is 100	is 1000	is 10,000

the intermediate numbers and divisions follow in common notation; for between

				Subdivisions.
1 and	10	the prime divisions are	Units	Tenths.
10 and	100	„	Tens	Units.
100 and	1,000	„	Hundreds	Tens.

It is here apparent that the lines *repeat*, at pleasure, and are therefore infinite, either in the ascending or descending scale. Had the numbering on the Rule been made absolute, its operations would have been limited, and valueless for questions above or below such stated quantities; the line D reads and repeats precisely in the same manner. It should be particularly noted when the squares and roots of numbers are in question, the true relation of C and D lines must be considered; thus, 2 on D is the square root of 4 on C, as 20 is of 400, *but not of* 40, which must be sought under the figure 4 on the right half of C, below which we find the correct square root of $40 = 6.32$.

C 1	4	40 right hand 4 on C
D 1	2	6.32 square root of 40

5th. The facility afforded by the Slide Rule in complex operations is greatly owing to the use of tabulated constants, the results of previous calculation, by the use of which fixed numbers all *similar* questions may be instantly solved; in Slide Rule practice these constants are *called* "Gauge Points," and are denoted by the abbreviation G. P., and any number in the body of this book so indicated is the constant of the question; but as substances vary in gravity, materials in strength, and bodies in dimensions; gauge points have to be found for each, and, when found, the labour saved by using them is almost incredible; sometimes their derivation is obvious, at others involved, but there is no mystery about them, and the beginner is not compelled to trace the nature of each before employing it; this can be done at leisure.

6th. To avoid the tedium of long printed directions for each working, which seriously divert the attention from the operation, the formula for stating each question will now be explained. It has been shown that the pairs of lines have different offices, and that the *proper pair* must always be employed. Great care has been taken to notify these, so that when $\frac{A}{B}$ or $\frac{C}{D}$ appear or are prefixed, that pair only is to be used. We will now take a diagram or formula from the body of the book; next show its full meaning; and then the simple working of the question.

See Measure of Capacity, Page 25.

C Length in feet.	Answer in cubic yards.
D G P 5.196.	Mean square feet.

FULL MEANING.

C Set the given length in feet on C.	Find the answer in cube yards on C.
D To the gauge point 5.196 on D.	Above the M. square in feet on D.

SIMPLE WORKING.

Ex.: Required the cubic yards excavated, when the length is 100 feet, and the width and depth 8 feet each.

C 100	237 cube yards, answer
D 5.196	8

The formula is intended to show invariably the lines to be used, and the relative positions of the terms; *not* their *actual places* on the Rule, for the latter vary with their values.

It is necessary always to note how the terms of the question are placed, and to follow the reading of the statement, viz:—First, look what numbers (or constants if given) are to be set, and *after that*, for the result; as

A	1st. $\begin{cases} \text{Set ?} \\ \text{To ?} \end{cases}$	Then Below ?
B		Find answer.

These *positions* may vary, as

A Then Below ?		
B Find answer	1st. $\begin{cases} \text{Set ?} \\ \text{To ?} \end{cases}$	

but in every case the directions are clear, and by the formulæ under any head all similar questions may be unhesitatingly worked by merely substituting the given numbers or dimensions for those in the examples.

Before putting these directions into practice, it may be as well to recapitulate their chief points.

1st. That the Rule is numbered, and divided decimally.

2nd. That any value may be assigned to a number, if carried throughout that line.

3rd. That the proper lines $\frac{A}{B}$ or $\frac{C}{D}$ are always to be used, and the question correctly stated, as in the formula

4th. That the reading or valuing of the numbers be well understood.

5th. That the correct G.P. is selected, and used as prescribed.

6th. That the positions in the formula do not necessarily imply the actual *places* of the numbers on the Rule, but are intended to show the exact order of stating *any similar* question.

ABBREVIATIONS.

G. P.	Denotes "gauge point." If the number is given, it is to be employed; but if not given, the proper one is to be selected.
a	(a) affixed to any number in a formula, shows the answer, or place of the answer.
A 1 or B 1, &c.	Shows that 1 (unity) on that slide or line is to be set to some given number.
×	Multiplier, or the *place* of a multiplier, or the office of the gauge point used.
÷	Divisor, with the same uses.
A B or C D	Prefixed to any formula, shows which lines are to be employed.
B. Invt^d C. Invt^d	The Slide has sometimes to be *inverted*; and in any operation where this is required, the letters Invt^d indicate it, and must be attended to.

THE ARRANGEMENT OF THE SLIDE RULE,

Issued with " Weale s Scientific Series."

THIS, the simplest form of the Slide Rule, is strongly recommended by the author for all ordinary purposes of calculation ; the separation of the usual lines A B, C D, needlessly connected on the common Rule, removes the appearance of intricacy so puzzling to the beginner ; and every formula in this work is adapted to a *single line and slide,* viz. :—

Either $\dfrac{A}{B}$ or $\dfrac{C}{D}$

and in each case the letters prefixed to the Example show unhesitatingly which pair is to be employed.

The Rule in this form being used for calculations only, and not like the common rule—half tool and half instrument—the utmost care has been expended on its construction, and its accuracy guaranteed.

LESSONS FOR PRACTICE.

THE correct reading of the numbers and divisions being of primary importance, so as to be able *at sight* to find or assign any required value, practice in this is essentially necessary, and the following method will greatly assist the beginner; A, B, and C lines being *exactly alike*, it is sufficient to take one of them, say A.

Left Half.	*Right Half.*
A 1 2 3 4 5 6 7 8 9 10	20 30 40 50 60 70 80 90 100

In the first reading, omit all divisions.

Then repeat, introducing the chief divisions.

A 1. { Read units and *tenths* up to........... 10   Read tens and *units* up to................100

Repeat, noting any subdivisions, as halves or fifths.

A 1. { Read units, tenths, and any subdivisions ... 10   Read tens, units, and any subdivisions........100

Now repeat, trying to estimate $\frac{1}{4}$, $\frac{1}{2}$, $\frac{3}{4}$, where undivided.

A 1.   10 100

Commence the line at 100, and estimate subdivisions.

A 100. { The *divisions* are tens............1,000   The *divisions* are hundreds 10,000

Having practised these scales sufficiently, try and make out the following numbers correctly :—

Left Half.	*Right Half.*
A 1. 2·2 2·7 3·6 4·9 7·5	12 15 27 34 50 63 75
144 225 337 450 725	3550 4300 5750 7500 8750

D   Although D, a single line, is to be read, valued, and *repeated* precisely as A, B, and C, yet affording more space, the divisions and subdivisions are carried further.

To confirm the above practice, let the learner test his accuracy by some *printed* tables; and as $\frac{C}{D}$ form of themselves lines of squares and roots (No. 5), proceed leisurely, and compare the Rule with the book; having the actual figures to guide and check the reading, let him try carefully to *estimate* the reading of undivided spaces, seeing how closely he can approximate to the 1st, then the 2nd place of decimals.

If books are handy, open a Table of Diameters and Areas ; and to form the same on the Rule, move the slide C gently till number 7 on C coincides with number 3 on D, by which a Table of Areas and Diameters is instantly made.

C .... The areas of circles on C .... Set 7
D .... To diameters　　on D .... To 3

then proceed with the comparison, till the book is used as a check, and not wholly as a guide. There is no preliminary practice so good as this, as it gives the learner confidence in the Rule and in himself. The perception of the power placed within his reach soon gives interest to each task, for the matter contained in *folios of such printed tables* lies in the instrument, available to all who have pride and patience enough to master it.

PART I.

INSTRUMENTAL ARITHMETIC.

The Notes on Slide Rule Arithmetic will be found page 12; if mixed with the Formula and Examples, they might divert attention from the simple working.

------

1. MULTIPLICATION on A and B. (Note 1).

Ex.: Multiply 9 by 3 = 27.

A To 3   Find 27 answer
B Set 1 on the Slide   Above 9 on the Slide

2. DIVISION on A and B.

Ex.: Divide 27 by 9 = 3.

A Find 3 answer   To 27
B Above 1 on the Slide   Set 9

3. PROPORTION (Direct) on A and B. (Note 2).

Ex.: As 2 : 6 :: 12 : 36.

A To 6   Find 36 answer
B Set 2 on the Slide   Above 12

When several proportions are required.

Ex.: As 18 : 252 :: 3', 4, 5, and 6 shares.

A Find 42   56   70   84 answer   To 252
B Above 3   4   5   6   Set 18

4. PROPORTION (Inverse) on A with the Slide *inverted*. (Note 3).

Ex.: As 6 : 4 :: 12 : 2.

A   Find   2   answer   To 4
S. Invt^d Above 12   Set 6

5. SQUARES AND ROOTS OF NUMBERS on C and D.

C Set 1　　　　(C the Slide) is then a line of square numbers.
D To 1　　　　D line shows the square roots of all Nos. on C.

Ex.: Find the square roots of 4, 9, 25, and 36.

C Set 1　　　　4　　　9　　　25　　　36
D To 1　　　　2　　　3　　　5　　　6 answers.

6. CUBES AND ROOTS on C and D.　(Note 4).

To cube a number or dimensions—*Ex.*: Cube 4.

C Set 4　　　　　　　　Find 64 the cube
D To 1　　　　　　　　Above 4

The given number on C may be set to 1 *or* 10, as the Slide some-times over-runs the numbers on D.

Ex.: Required the cube root of 64.　(Invert the Slide to D).
　　Note 4.

(S I) Find 4 the root　　　　　　Set 64
D　　　4　　　　　　　　　　To 10

7. 4th POWER AND ROOTS on C and D.　Note 5.

Ex.: Find the 4th power of 3.

C 1st. { Find　9 its 2nd power.　　Find 81 the 4th power
D 1st. { Above 3　　　　　　　Then above 9

Ex.: Find the 4th root of 81.

C Then below 9　　　　　　　1st { Below 81.
D　　Find 3 the 4th root of 81　　　　{ Find　9 its sq. root.

8. MEAN PROPORTIONAL, or MEAN SQUARE of UNEQUAL SIDES on C and D.　See Note 6.

Ex.: Mean Square of 4 and 9.

C　Set 4 the less No.　　　　Below 9 the greater No.
D　To　4 its like No.　　　　Find　6 the mean.

9. Two MEAN PROPORTIONALS, a double setting required.

Ex. Find two mean proportionals to 2 and 16.

1st, with Slide inverted.

(S. Invt⁴) Set 16		4 ⎫	are coincident numbers and the
D	To 2	4 ⎭	1st proportional.

2nd with Slide rectified.

C Set 4, 1st proportional	Below 16.
D To 4, its like No.	Find 8 the 2nd proportional.

Answer, 2 4 8 16.

10. To MULTIPLY SQUARED NUMBERS.

In effect the same as Cubing. See No. 6 and Note 7.

Ex.: Multiply 3.5^2 by 4.

C Set 4	Find 49 the answer.
D To 1	Above 3.5

11. To DIVIDE SQUARED NUMBERS C and D.

Ex. Divide 6^2 by 4.

C Set 4	Find 9 the answer.
D To 4	Above 6.

12. To DIVIDE BY A SQUARED NUMBER, C and D.

Ex.: Divide 144 by 3^2.

C Find 16 answer	Set 144
D Above 1	To 3

DECIMAL AND COMMON FRACTIONS.

13. DECIMAL EQUIVALENTS, OR A and B.

Ex.: Convert the fraction ⅝ to a decimal.

A Find the .625 the dec. equivalent		Set 5 ⎫	or with any given
B Above	1 on the Slide	To 8 ⎭	fraction on the rule.

14. DECIMALS TO COMMON FRACTIONS.

Ex. : Convert the decimal .625 to a common fraction.

A Set .625   Find $\left\{ \dfrac{5}{8} \right.$ (Note 8).
B To 1

15. RECIPROCALS on A and B.

Ex.: Find a divisor equal to the multiplier 4.

A Below 1   To 4 ×
B Find .25 ÷ answer   Set 1.

Ex. : Find a multiplier equal to the divisor 5.

A Below 1.   To 5 ÷
B Find .2 × answer   Set 1

16. COMMON FRACTIONS, valued in any denominations on A and B.

Ex. : Find the value of $\frac{3}{8}$ of 20 shillings.

A Find 7.5 shillings   To 20 shillings
B Above 3   the numerator   Set 8 the denominator

Any fraction may be as easily reduced, as $\frac{9}{15} = \frac{3}{5}$ of 37.

A 22.5 answer   37.
B 3   5.

17. TO FIND THE VALUE OF DECIMAL PARTS OF ANY INTEGER.

Read the *whole line* A as a scale of decimal parts, thus :—

Beginning   Centre.   Tenths.   Right.
A .01   .02   .03 &c. to .1   .2   .3   .4 &c. to 1. whole

Any denomination being placed under 1, on the *right* of the Scale A. the decimal values of any *parts* on B are shown on A, or on B, find the value in parts of any decimal on A. (Note 9.)

NOTES ON INSTRUMENTAL ARITHMETIC.

NOTE 1.—*Multiplication.*—When 1 on the Slide is set to any number on A, *all* numbers assume that ratio, and the product of ANY number on B may be found on A without shifting the Slide.

Note 2.—*Proportion Direct.*—If more requires more, or less requires less, the question is one of Direct Proportion. The statement on the Rule may be varied thus :—

Let	a	b	c	d				
Represent	2	4	6	12 .	:	: :		:

Then, by Slide Rule—

$$x = \frac{b \cdot c}{d} \qquad \begin{array}{ll} \text{A } 2\ x \\ \text{B } 6\ c \end{array} \quad \begin{array}{l} 4\ b \\ 12\ d \end{array} \quad \text{or} \quad \begin{array}{ll} c & 6 \\ x & 2 \end{array} \quad \begin{array}{l} 12\ d \\ 4\ b \end{array}$$

$$x = \frac{a \cdot d}{c} \qquad \begin{array}{ll} \text{A } 4\ x \\ \text{B } 12\ d \end{array} \quad \begin{array}{l} 2\ a \\ 6\ c \end{array} \quad \text{or} \quad \begin{array}{ll} d & 12 \\ x & 4 \end{array} \quad \begin{array}{l} 6\ c \\ 2\ a \end{array}$$

$$x = \frac{a \cdot d}{b} \qquad \begin{array}{ll} \text{A } 6\ x \\ \text{B } 2\ a \end{array} \quad \begin{array}{l} 12\ d \\ 4\ b \end{array} \quad \text{or} \quad \begin{array}{ll} a & 2 \\ x & 6 \end{array} \quad \begin{array}{l} 4\ b \\ 12\ d \end{array}$$

$$x = \frac{b \cdot c}{a} \qquad \begin{array}{ll} \text{A } 12\ x \\ \text{B } 4\ b \end{array} \quad \begin{array}{l} 6\ c \\ 2\ a \end{array} \quad \text{or} \quad \begin{array}{ll} b & 4 \\ x & 12 \end{array} \quad \begin{array}{l} 2\ a \\ 6\ c \end{array}$$

In Direct Proportion, the multipliers never appear on the same line, and are always in opposition,

$$\text{As } \frac{\text{A}}{\text{B}} \qquad \begin{array}{c} \text{Answer} \\ \times \end{array} \quad \begin{array}{c} \times \\ \div \end{array} \quad \text{or} \quad \begin{array}{c} \times \\ \div \end{array} \quad \begin{array}{c} \text{Answer} \\ \times \end{array}$$

Note 3.—*Proportion Inverse.*—If more requires less, or less requires more, the question is in Inverse Proportion.

There are two methods of stating such questions. I believe I recommend the simplest, in *reversing* the Slide, for the lines A B are thus changed from Direct to *Inverse* Proportion, when the statement at once becomes clear.

Ex.: As 6 : 4 :: 12 : 2.

A	Set 6	Below 12.
(S. Inv^d)	To 4	Find 2 Answer.

It is only necessary for observation, but useful to note, that in all Inverse questions, if correctly stated, the products of the vertical numbers on the rule are equal.

$$\begin{array}{c} 6 \\ 4 \\ \hline 24 \end{array} \qquad = \qquad \begin{array}{c} 12 \\ 2 \\ \hline 24 \end{array}$$

Note 4.—*Cube Root.*—This tedious arithmetical process would be performed on the Slide Rule by mere inspection, the same as squares and roots, were C a *triple* line to D single; in fact, any roots and powers might be found by increasing the sections of C to the required power (the present double radius gives the 2nd, and by No. 7, the 4th, and 8th). In the absence of an arrangement, which would affect the simple form—squares and roots being sufficient for all **general** purposes—a little ingenuity must be substituted to find cube roots; we therefore *reverse the Slide*, and, setting the given number to 1 or 10 on D, have to look for the numbers or divisions exactly coinciding, which is the required cube root; as in the example—

$$64 \sqrt[3]{}$$

S. Inv^d Find the coincident $\begin{cases} 4 \\ 4 \end{cases}$ $\begin{cases} \text{the cube root} \\ \text{of 64.} \end{cases}$ 1st $\begin{cases} \text{Set 64.} \\ \text{To 10.} \end{cases}$
D   Numbers ......

In this case it is easy and evident that no other similar numbers meet; but, sometimes, owing to the differing scales of C and D, and to one being inverted, a little patience is needed, to mark the intersecting numbers or divisions on $\Big\}$ C Ivt^d the only case in Slide Rule D, practice demanding it, but still wonderfully simple compared with arithmetic.

Note 5.—*4th Power and Root.*—To those acquainted with Evolution, the example given will suffice, for C and D forming lines of squares and roots, we can find the square root of the square root; and, consequently, the 4th power or root of any number by inspection.

Note 6.—*Mean Proportional, or Mean Square.*—This rule is important in its application to cubed work, for, *unless* the area of the section is found, the true mean square must be used in multiplying by the length, or serious error occurs.   For instance— .

$$4 \times 9 = 36 \sqrt{} \quad = 6, \quad \text{the true mean.}$$
$$\text{whereas } 4 + 9 = 13 \div 2 \quad = 6.5, \text{ the arithmetical mean.}$$

Wherever the mean square is demanded in this work, it must be found by the easy method given page 10.

The effect of setting the 4 on C   Is to project the 9 on C
To its like No.   4 on D   To their sq. root 6 on D

Note 7.—*Multiplying Squared Numbers.*—This is the same as in cubing; for the line D being already, by its proportions, equal to the *square* of its numbers on C,

as   C 1   4   9   &c., &c.
     D 1   2   3

and the setting of any number on C ?
                                    to D 1,

being to multiply *all* numbers on D by that number. If it represents the length of a solid, we have

> Squared-up section × length, or the cube contents on C.

Note 8.—*Converting Decimal and Common Fractions.*—When unity (1 on B) is set to any decimal on A,

then A represents the numerators ⎫ of all corresponding
and B   „   „   denominators ⎬ common fractions,

from which the lowest term can be selected by inspection. For example—

A   To .125   .25   .375
B   Set   1.   gives $\frac{1}{8}$ $\frac{1}{1}$ gives $\frac{1}{4}$ $\frac{.375}{1}$ gives $\frac{3}{8}$, &c.

But should no unit divisions coincide, the fraction may be found among the tens.

$$Ex.: \quad \begin{array}{l} A \\ B \end{array} \quad \begin{array}{l} \text{To } .766 \\ \text{Set } \; 1 \end{array} \; = \; \frac{23}{30}$$

Note 9.—*Decimal Values.*—The construction of the Slide Rule, necessitating the reduction of all statements and readings to decimal expressions, facility in this operation is essential; but the examples given are so clear, and the methods so speedy and simple, even with the complex divisions and subdivisions of our standards, that a little practice removes every difficulty. It is recommended to compare the readings of the instrument with some *printed* decimal tables. A correct knowledge of the scale of divisions is, at the same time, acquired; and such lessons will be found to have many advantages over unchecked readings.

Ex. : Place *in succession*, under 1, on the right of A, the denominators of a £, a yard, a rod, an acre, or of *any* integer, and form decimal scale and value.

                     under
A Read Line of decimal parts on A÷1 on the right.
                     Set

B			÷20 shillings	== 1 £.
B	Only one denomination at a time.	Above any Divisions of the Integer	36 inches	1 yard.
B			272 feet	1 rod.
B			160 rods	1 acre.
B			12 pence	1 shilling.
B			12 inches	1 foot.

The different processes in Slide Rule Arithmetic will be found applied in the Parts which follow; and reference made to these notes, where necessary, to explain any operation more fully than the general formula would permit.

PART II.

MENSURATION.

———

Questions in Mensuration are solved with remarkable facility by the Slide Rule; for not only are the tedious processes in finding the contents of solids avoided altogether, but, in the majority of cases, given (see Note 5) tabular results of great value are obtained. For greater convenience of reference, this subject is divided into— 1st, Linear; 2nd, Superficial; and 3rd, Solid Measurement.

———

1st. LINEAR MENSURATION.

———

1. Diagonals of Squares on A B :

A	Find the diagonals of all squares	1st {	Set 99
B	Above the sides do.		To 70

Or on C and D :

C	Set 1	Below 2
D	To side of any square	Find diagonal

Ex.: The side of a square being 3.5, required its diagonal.

C	Set 1	Below 2
D	To 3.5	Find 4.95 diagonal

2. Diagonals of Cubes on C and D :

C	Set 1	Below 2	Below 3
D	To side of *any* cube	Find diag. of face	Find diag. of cube

Ex.: The side of a cube being 3.5, required the diagonals of its face and solid.

C	Set 1	Below 2	Below 3
D	To 3.5	Find 4.95 diag. of face	Find 6.06 diag. of cube

3. One Mean Proportional (on C and D).

C	Set the least number	Below the greater number
D	To the least number	Find the mean proportional

Ex.: Find a mean proportional to 4, and 9.

| C | Set 4 | Below 9 |
| D | To 4 | Find 6 the mean |

4. TWO MEAN PROPORTIONALS (on C D).

| (Inv^td Slide). | Set the greater No. | * | Observe where any numbers |
| D | To the least | * | or divisions coincide, *they will be 1st proportional.* |

Then rectify the Slide.

| C | Set * 1st proportional | Below the greater Number |
| D | To * 1st proportional | Find the 2nd proportional required |

Ex.: Find two mean proportionals to 2 and 16.

| Slide Inv^td | Set 16 greater No. | 4 will be found coincident and |
| D | To 2 least No. | 4 is the 1st proportional |

Then—

| C | Set 4 | Below 16 |
| D | To 4 | Find 8 the 2nd proportional |

Answer—2 . 4 . 8 . 16.

5. SIDES OF SQUARES OF EQUAL AREA (on A and B).

A B	Tabular {	Find sides of squares of equal area	Set 39
		Above diameters of circles	To 44
A B	Tabular {	Find sides of squares of equal area	Set 11
		Above circumferences of circles	To 39

6. SIDES OF SQUARES TO BE INSCRIBED (on A and B).

A B	Tabular {	Find sides of squares to be inscribed	Set 53
		Above diameters of circles	To 75
A B	Tabular {	Find sides of squares to be inscribed	Set 45
		Above circumferences of circles	To 200

7. DIAMETERS OF CIRCULAR SEGMENTS (on A and B).

| A | Set the versed sine | Below the ½ chord |
| B | To the ½ chord | Find the supplement, which, added to the v. sine = diameter |

Ex.: The chord of a segment being 48, and the versed sine 18, required the diameter of the circle.

A   Set v. sine = 18   Below 24 = ½ chord
B   To ½ chord = 24   Find   32 the Supplement

$$\left.\begin{array}{cc} \text{Supplement (add) to v. sine} \\ 32 \qquad\qquad + \qquad\qquad 18 \end{array}\right\} = 50 \text{ the diameter required}$$

8. LENGTH OF ARC (*) DEGREES (on A and B).

Note.—Find the diameter by the last rule if necessary.

A   To 115   Below any given degrees
B   Set the diameter   Find the length of arc

9. CIRCUMFERENCE AND DIAMETER OF CIRCLE (on A and B).

A   Set 22   Line of circumferences
B   To   7   Line of diameters.

10. ELLIPSE CIRCUMFERENCE (on A and B).

A   Set 70   Find the circumference of any ellipse
B   To 45   Above the *sum* of the two diameters

11. ISOMETRICAL ELLIPSE (on C and D).

C   Set 1   Below 2.   Below 3.
D   To the Minor Axis   Find Isomet. Diam^r.   Find the Major Axis.

12. RIGHT-ANGLED TRIANGLES (on C and D).

To find the hypothenuse—

C   Set 1 Find its square.   Find its square (add) $\left\{\begin{array}{l}\text{and below the}\\ \text{sum of squares}\end{array}\right.$

D   To 1 Above the 1st leg   Above the 2nd leg $\left\{\begin{array}{l}\text{Find the hypothe-}\\ \text{nuse on D.}\end{array}\right.$

   Ex.: The sides of R. A. triangle being 4 and 6, required the hypothenuse.

C   Set 1 Find   16 +   36 =   52 below which find
D   To 1 Above 1st leg 4   2nd leg 6   7.21 the hypothenuse.

   To Find either leg.

     The hypothenuse added to one leg = Sum *
              Then—

C   Set * sum   Below hypothenuse *less the same leg*.
D   To * sum   Find the required leg.

13. OTHER TRIANGLES (on A and B).

Note.—By the use of the natural sines, &c., the sides and angles of triangles may be found on A and B; space cannot be given for more than the sines and differences of *full degrees*, to show their use, or for approximate working. Lines of sines, tangents, &c., can be laid down on a Slide, and by their aid and the line A, the necessity for any tables is avoided, as all trigonometrical questions that are solved by using logs and log sines are answered on the Rule.

TABLE of Natural Sines to full Degrees, with differences from which 10′ 20′ &c., may be found.

Deg. N.S.	Dif.	Deg. N.S.	Dif.	Deg. N.S.	Dif.	Deg. N.S	Dif.
1 .017		23 .391		45 .707		69 .933	
2 .035		24 .407		46 .719		70 .939	.006
3 .052		25 .423	.016	47 .731	.012	71 .945	
4 .070		26 .438		48 .743		72—.951	
5 .087		27 .454		49—.755		73 .956	.005
6 .104		28 .469		50 .766		74 .961	
7 .122		29—.485		51 .777	.011	75—.966	
8 .139	difference .017	30 .5		52—.788		76 .970	.004
9 .156		31 .515	.015	53 .798		77 .974	
10 .174		32 .530		54 .809		78—.978	
11 .191		33—.545		55 .819	.010	79 .981	
12 .208		34 .559		56 .829		80 .984	.003
13 .225		35 .573		57 .838		81 .987	
14 .242		36 .588	.014	58—.848		82—.990	
15 .259		37 .602		59 .857	.009	83 .992	.002
16 .276		38 .616		60 .866		84 .994	
17 .292		39 .629		61—.875		85—.996	
18 .309		40—.643		62 .883	.008	86 .997	
19 .325		41 .656		63 .891		87 .998	.001
20 .342		42 .669	.013	64—.899		88—.999	
21 .358		43 .682		65 .906		89 .999	
22—.375		44—.695		66 .913	.007	90 1. —	
				67 .920			
				68—.927			

Ex.: The side of a triangle (whose opposite angle is 40° sine 643 from Table) equals 800 yards; required the side, the angle being 30° sine .5 in table.

A Find 622 yards required side

B Above ·5 the sine of 30″

To 800 the yards in given side.

Set ·643 the sine of 40°

14. REGULAR POLYGONS Linear proportions (on A and B).

			Sides.	3.	4.	5.	6.	7.	8.	9.	10.	11.	12.
Find Radius.............	On A	Set		56	70	74	1	60	98	22	89	80	85
Above Len. of Side......	On B	To		97	99	87	1	62	75	15	55	45	44
Find Perpendicular	On A	Set		9	1	40	26	25	40	40	40	29	28
Above Len. of Side......	On B	To		31	2	58	30	23	33	29	26	17	15
Find Perpendicular ...	On A	Set		1	20	21	26	27	25	33	20	25	28
Above Radius	On B	To		2	28	26	30	30	27	35	21	26	29

Note.—When the constants for any polygon are set on A and B, the proportions named in the margin appear in tabular form on their respective lines for all figures of that kind.

Ex.: It is required to form a table of side and radius for octagons.

A is then a line of radii
B „ „ sides

1st { Set 98 on A { Constants for
{ To 75 on B { octagons.

15. DIVIDING LINES INTO EQUAL PARTS on (A and B).

A Find length of part required   To the whole length
B Above 1   Set the No. of parts

2ND. MENSURATION OF AREAS.

1. SQUARE SUPERFICES (on A and B).

When dimensions in Feet × Feet, the answer in *square feet*.
A To 1   Below the breadth
B Set length feet   Find square feet super

When dimensions in Feet × Inches, the answer in sq. feet.
A To 12   Below the breadth in inches
B Set length feet   Find sq. feet super

When dimensions in Inches × Inches, the answer in square feet.
A To 144   Below breadth in inches
B Set length inches   Find sq. feet super

When dimensions in Yards × Yards, the answer in *square yards*.
A To 1   Below the breadth in yards
B Set length in yards   Find sq. yards super

When dimensions in Yards × Feet, the answer in square yards.
A To 3
B Set length in yards

Below the breadth in feet
Find sq. yards super

When dimensions in Feet × Feet, the answer in square yards.
A To 9
B Set length in feet

Below the breadth in feet
Find sq. yards super

When dimensions in Chains × Chains, the answer in *acres*.
A To 10
B Set length in chains

Below the breadth in chains
Find acres area

When dimensions in Perches × Perches, the answer in acres.
A To 160
B Set length in perches

Below the breadth in perches
Find acres area

When dimensions in Yards × Yards, answer in acres.
A To 4840
B Set length in yards

Below the breadth in yards
Find acres area

When dimensions in Yards × Yards, answer in perches.
A To 30.25
B Set length in yards

Below breadth in yards
Find perches area

Ex.: A sheet of plate glass is 7 feet long, and 34½ inches wide—required the sq. feet.

A To 12
B Set 7 feet long

Below 34 inches wide
Find 20½ sq. feet super

2. TRIANGLES (on A and B).

A Set length of bas
B To 2

Find the area
Above the perpendicular

3. REGULAR POLYGONS (on C and D).

		Sides.	3.	4.	5.	6.	7.	8.	9.	10.	11.	12.	16.	20.	24.
Find Areas ...	On C	Set	21	9	43	65	58	310	500	490	150	100	500	500	410
Above Sides .	On D	To	7	3	5	5	4	8	9	8	4	3	5	4	3

Ex.: Required the area of a Pentagon, the side being 4 feet (and form a table of areas of Pentagons).

C Find 27.5 area	Set 43 ⎱ Constants for Penta-
D Above 4 the given sides	To 5* ⎰ gons.

4. TRAPEZOIDS (on A and B).

A Set the sum of Parallel sides	Find super. contents
B To 2	Above the breadth

5. *Circular Areas.*

6. DIAMETERS AND AREAS (on C and D).

C Set 7	Line of areas on C
D To 3	Line of diameters on D

Ex. Find the Areas to Diameters 5, 6, and 7.

C Set 7	Find 19.6, 28.3, 38.5 areas required
D To 3	Above 5, 6, 7 diameters given

7. DIAMETERS IN INCHES, *Areas in Square Feet* (on C and D).

C Set 11	Find areas in square feet on C
D To 45	Above any diameter in *inches* on D

8. CIRCUMFERENCES AND AREAS (on C and D).

C Set 23	Line of areas on C
D To 17	Line of circumferences on D

(Or on A and B).

A Set the radius	Find the area on A
B To 2	Above the circumference on B

9. AREAS OF CIRCLES AND INSCRIBED SQUARES (on A and B).

A Set 11	Find areas of circles on A
B To 7	Above areas of inscribed squares

Nos. 6 to 9, yield tabular results.

10. AREAS OF SECTORS OF CIRCLES (n) DEGREES (on C and D).

C Set 465	Below any given No. of Degrees
D To diameter of circle	Find square root of area

Close the Slide, and above the square root on D, find the square = area of sector on C.

(Or on A and B).

A Set the radius of circle   Find area of sector required
B To 2   Above the length of arc

Example of No. 10. The diameter of a circle is 20, the degrees of the sector 15 ; required the area.

C Set 465   Below 15°
D To   20   Find 3.62 the square root of area
                                              = 13,09 area

11. SQUARE AND CIRCULAR AREAS COMPARED (on A and B).

A Set 70   Line of circular feet, inches or measures
B To 55   Line of square   „   „   „

Note.—The circular foot contains 113.0977 square inches.
     The square foot contains 183.346 circular inches.

12. ELLIPTICAL AREAS (on A and B).

A Set 55   Find area of ellipse on A
B To 70   Above the *product* of its diameters on B

Ex.: required the area of an ellipse, the diameters being 15 and 20 = 300.

A   55   235.6 area
B   70   300

                          or

A Set long. diameter   Find area
B To 1.273   Above short diameter

13. PARABOLIC AREAS (on A and B).

A Set the base   Find the area
B To 1.5   Above the altitude

14. CONICAL SURFACES (on A and B).

A Set the circumference of base   Find the curve surface
B To   2   Above the slant height

Or—

A Set the diameter   Find the curve surface
B To   .64   Above the slant height

Ex. of No. 14. The diameter of the base of a cone is 5 inches, its slant height 18 inches ; required the curve surface.

A 5. Diameter   141.3 Curve surface required
B .64   18 Slant height

15. CONVEX OF SPHERES (on C and D).

C Set 50   Find the convex surface
D . To 4   Above the diameter of any sphere

Ex.: The diameter of a sphere is 8 ; required its convex surface.

C 50   201 convex surface
D 4.   8 diameter

16. CONVEX OF SPHERICAL SEGMENTS (on A and B).

A Set height of segment   Find convex of segment
B To   1   Above circumference of sphere

Ex.: The circumference of a sphere being 25.1, and the height of a segment of it being 2, required the convex surface.

A 2 H   50.2 convex surface, answer
B 1   25.1 circumference

17. SURFACE OF REGULAR SOLIDS (on C and D).

Number of facets	4	6	8	12	20
Find the whole surface on C Set	85	24	170	330	78
Above the edge of facet on D To	7	2	7	4	3

Ex.: Required the surface of a regular solid of 8 sides, the edges of each facet being 4 inches.

C * 170   55.5 inches surface
D   7   4 inch edge

18. CURVE SURFACE OF CYLINDERS (on A and B).

| A | Set the diameter | Find the curve surface |
| B | To   .32 | Above the height |

Or—

| A | Set the circumference | Find the curve surface |
| B | To   1 | Above the height |

Ex.: The diameter of a cylinder is 8 inches, and its height is 12 inches; required its curve surface.

| A | 8 inches | 301.6 inches curve surface |
| B | .32 | 12   height |

Note.—The greatest cube content of a cylinder bounded by the least surface is when the diameter is made equal to the required cube contents multiplied by 2.5465, and the cube root of the product taken; its depth being half such diameter.

3RD. MENSURATION OF SOLIDS.

---

1. CUBICAL FORMS (on C. and D).

When the cube contents are required in the *same measure* as that in which the dimensions were taken.

| C | Set the length | Find the cube contents |
| D | To 1 | Above the mean square of side |

Ex.: The length of a shaft being 30 feet. and its sides 5 × 4 feet; required the cube contents.

| C | 30 L | 600 Cube feet content |
| D | 1 | 4.47 Mean square of 5 × by 4 |

A simple case has been given, but the importance of finding the true mean square in cubical measurements must never be overlooked. (See Part 1, No. 8).

When the cubical capacity is required in various measures, as cube yards, gallons, &c., and the dimensions are taken sometimes in feet, at others in inches, the proper gauge points, which are here given, must be used; if examined, they will be found to be the square

c

roots of the contents of the integer of measure used in the question to be answered.

2. Dimensions being all in feet, and the answer in CUBE YARDS:

C Set the feet length   Find the cube yards contents
D To 5.196   Above the M. sq. feet of sides

3. Dimensions being all in feet, and the answer in GALLONS:

C Set the feet length   Find the gallons content
D To .4   Above the M. sq. feet of sides

4. Dimensions being all in feet, and the answer in BUSHELS:

C Set the feet length   Find the bushels contents
D To 1.131   Above the M. sq. feet of sides

5. Dimensions being all in inches, and the answer in CUBE FEET:

C Set the inches length   Find the cube feet content
D To 41.57   Above the M. sq. inches of sides

6. Dimensions being all in inches, and the answer in GALLONS:

C Set the inches length   Find the gallons contents
D To 16.65   Above the M. sq. inches in side

7. When the length is given in feet and the sides in INCHES:

C Set the feet length   Find the cube feet contents
D T. 12   Above the M. sq. in inches

Example of No. 7. A cistern is 40 inches long, its sides 30 × 25 inches, required its contents in gallons.

C 40 inches length   108.2 gallons contents
D 16.65   27.4 in M. sq. of 30 × 35

The Gauge point $16.65 = \sqrt{277.27}$ cubic inches in a Gallon.

The cubing of unequal sided bodies may be effected as follows, but the operation is less simple than the method given; and is inserted merely to compare the different workings; the employment of 4 lines of the Rule when 2 will effect the same purpose is always objectionable, whatever facility practice may give in their use.

A   2nd   3rd $\left\{\begin{array}{l}\text{Then below the length on A} \\ \text{Find cube ft. contents on B}\end{array}\right.$

B Set one side of section on B

(C Inverted) to other side „ on C   1st $\left\{\begin{array}{l}\text{Set the divisor*} \\ \text{To 10 on D}\end{array}\right.$

D

When Section	Length	*Divisors	
In. × In.	Inches.	1728	
In. × In.	Feet.	144	Answer
In. × Ft.	Feet.	12	in
Ft. × Ft.	Ft.		Cubic

Or any dimensions when the Section and $\left.\begin{array}{c} \\ \text{use 1}\end{array}\right\}$ Feet.
Length are in the same measure, and the
answer of the same kind.

8. CUBICAL PROPORTIONS.

A cubic foot contains—

1728 Cubic	inches	1.	Cube 1.		
2200 Cylindrical	„	.7854	1.273	Cylr. 1.	
3300 Spherical	„	.5236	1.909	2. Sphere 1	
6600 Conical	„	.2618	3.819	3.	2

And to reduce or compare Cubic with other Measures (on A and B).

A Set 55   Line of cubic   measures on A.
B To 70   „   cylindrical   „   on B.

A Set 42   „   cubic   „   on A.
B To 80   „   spherical   „   on B.

A Set 21   „   cubic   „   on A.
B To 80   „   conical   „   on B.

9. OF CYLINDRICAL FORMS (on C and D).

When the cube contents are required in the same measure as that in which the dimensions were given.

GENERAL RULE.

C Set the length   Find cube contents
D To 1.128   Above the mean diameter

o 2

Or, with circumference:

C Set the length	Find cube contents
D To 3.54	Above the circumference

Ex.: A cylinder is 6 feet long and 4 ft. 6 in. diameter, required its contents in cube feet.

C Set 6 feet	Find 95.4 cube feet contents
D To 1,128	Above 4.5 feet diameter

To Find the Capacity in Various Measures.

10. Dimensions being all in feet, and the answer in CUBE YARDS:

C Set feet length	Find the contents in cube yards
D To 5.86	Above mean diameter in feet

11. Dimensions being all in feet, and the answer in GALLONS:

C Set feet length	Find the contents in gallons
D To .451	Above mean diameter in feet

12. Dimensions being all in feet, and the answer in BUSHELS:

C Set feet length	Find the contents in bushels
D To 1.277	Above the mean diameter in feet

13. Dimensions being all in inches, and the answer in CUBE FEET:

C Set inches length	Find contents in cube feet
D To 46.9	Above mean diameter in inches

14. Dimensions being all in inches, and the answer in GALLONS:

C Set inches length	Find contents in gallons
D To 18.79	Above mean diameter in inches

Ex.: A cask being 40 inches long and 30 inches mean diameter, required the contents in gallons.

C 40 inches length	Find 102.3 gallons content
D *18.79	30 inches mean diameter

Note.—The gauge point *18.79 is the square root of 353, the number of cylindrical inches in a gallon.

15. When the length is given in feet and diameter in Inches:

C Set the feet length   Find the cube feet contents
D To 1.35   Above the M. diameter in inches

Or, with circumference:

C Set the feet length   Find the cube feet contents
D To 4.25   Above the circumference in inches

Note.—Cylindrical gauge points for Diameter × 3.14 give gauge
points for circumference.

16. CYLINDRICAL PROPORTIONS.

A cylindrical foot contains—

1357.17 Cubic inches.	Cylinder 1.
2591.8 Spherical „	Sphere 2.
5183.6 Conical „	Cone 3.

17. To REDUCE AND COMPARE CYLINDRICAL WITH OTHER MEASURES (on A and B).

A Set 70   Line of cylindrical measures on A
B To 55   „   cubical measures on B

A Set 22   „   cylindrical measures on A
B To 33   „   spherical measures on B

A Set 22.   „   cylindrical measures on A
B To 66   „   conical measures on B

18. CONTENTS OF CYLINDERS 1 FOOT IN DEPTH (C and D).

Diameters in feet.

C Set 42   Find cubic yards at 1 foot deep
D To 38   Above diameters in feet

C Set 7   Find cubic feet at 1 foot deep
D To 3   Above diameter in feet

C Set 490   Set gallons at 1 foot deep
D To 10   Above diameter in feet

19. Diameters in INCHES.

C Set 34	Weight of water in lbs. at 1 foot deep
D To 10	Above diameter of cylinder in inches
C Set 11	Cubic feet at 1 foot deep
D To 45	Above diameter of cylinder in inches

———

20. SPHERICAL FORMS (on C and D).

When the cube contents are required in the same measures as that in which the dimensions were taken.

C Set the diameter	Find the cube contents
D To 1.382	Above the diameter

Or, for circumference:

C Set the diameter	Find the cube contents
D To 4.34	Above the circumference

Note.—The diameter of a sphere is taken twice, as it answers for length.

Ex.: The diameter of a sphere being 4 feet, required its cube contents.

C 4 diameter	33.5 cube feet content.
D 1.382	4 diameter.

When the cube contents are required in various measures, use the prescribed gauge points.

21. Dimensions being all in feet, and the answer in CUBE YARDS:

C Set the feet diameter	Find the cube yards contents
D To 7.18	Above the diameter in feet

22. Dimensions being all in feet, and the answer in GALLONS:

C Set the feet diameter	Find the gallons contents
D To .553	Above the diameter in feet

23. Dimensions being all in feet, and the answer in BUSHELS:

C Set the feet diameter   Find the bushels contents
D To 1.563   Above the diameter in feet

24 Dimensions being all in inches, and the answer in CUBE FEET:

C Set the inches diameter   Find the cube feet contents
D To 57.44   Above the diameter in inches

25. Dimensions being all in inches, and the answer in GALLONS:

C Set the inches diameter   Find the gallons contents
D To 23.02   Above the diameter in inches

Example of No. 22. The diameter of a sphere being 3 feet 6 inches, required its contents in gallons.

C Set 3.5 diameter   140 gallons contents
D To .553   3.5 diameter.

26. SPHERICAL PROPORTIONS.

A Spherical foot contains
904.78 Cubical inches
1152 Cylindrical „
3456 Conical „

And to reduce or compare Spherical with other Measures (on A and B).

A Set 80   Line of spherical measures on A
B To 42   „ cubical „ on B

A Set 33   „ spherical „ on A
B To 22   „ cylindrical „ on B

A Set 33   „ spherical „ on A
B To 66   „ conical. „ on B

27. SPHEROIDS (on C and D.)
Dimensions in same MEASURES.

C Set the fixed axis.   Find the cube contents
D To 1.302   Above the revolving axis

28. REGULAR CONES, RIGHT OR OBLIQUE (on C and D).

C Set the altitude *   Find the cube contents
D To 1.954   Above the diameter of base

Or, for circumference.

C Set the altitude   Find the cube contents
D To 6.14   Above the circumference of base

* Not the slant height.

29. FRUSTUMS OF REGULAR CONES (on A and B).

A { Set the sum of the diameters } Find the cube contents
 { squared, less their product }
B   To 3.82   Above the altitude

30. CONES, REGULAR OR IRREGULAR (on A and B).

A Set the area of base   Find the cubic contents
B To 3   Above the altitude

31. FRUSTUM OF IRREGULAR CONES.

A { Set the sum of areas added } Find the cube contents
 { to their mean proportional }
B   To 3   Above the altitude

32. PROPORTION OF CONES (on A and B).

A conical foot contains :

452.39 Cubic inches
576. Cylindrical inches
864. Spherical inches

To reduce or compare

A Set 80   Line of conical measures
B To 21   ,,   cubic ,,

A Set 66   ,, conical ,,
B To 22   ,, cylindrical ,,

A Set 66   ,, conical ,,
B To 33   ,, spherical ,,

33. SQUARE PYRAMIDS (on C and D).

C Set the altitude   Find the cube contents
D To   1·732   Above the side of base

34. FRUSTUM OF SQUARE PYRAMIDS (on A and B).

A Set { The *sum* of the sides squared }   Find the cube contents
      { less their product }

B To   3   Above the altitude

35. PYRAMIDS, REGULAR OR IRREGULAR (on A and B).

A Set the *area* of base   Find the cube contents
B To   3   Above the altitude

36. FRUSTUM OF REGULAR OR IRREGULAR PYRAMID (on A and B).

A Set { The *sum* of areas added }   Find the cube contents
      { to their mean proportional }

B To   3   Above the altitude

37. PARABOLOID (on A and B).

A Set the altitude   Find the cube contents
B To   2   Above the *area* of base

(On C and D).

C Set the altitude   Find the cube contents
D To   1·596   Above the *diamr.* of base

38. FRUSTUM OF PARABOLOID (on A and B).

A Set { The *sum* of the }   Find the cube contents
      { areas of the ends }

B To   2   Above the altitude.

39. CUBE CONTENTS OF REGULAR SOLIDS (on C and D).

No. of Facets.	Gauge Points.	
4	2·91	The length of the side, or
6	1·	edge of the facet given,
8	1·46	to find the cube con-
12	·36	tents
20	·698	

[See next page for Examples.]

B 3

C Set the length of edge   Find the cube contents
D To the G P   Above the length of edge

 Ex.: An eight-sided solid, with 6-inch edges, required the cube contents.

C   6-inch edge   103.3 inches cube contents
D   1·46 for 8 sides   6-inch edge

40. PRISMS, REGULAR (on C and D).

No. of Sides.	Guage Points.	
3	1.523	The length of the prism
4	1.	or shaft, and the mea-
5	.762	sure of the side of the
6	.62	section being given, to
7	.523	find the cube content
8	* .456	
9	.4	
10	.36	
11	.326	
12	.298	

C Set the length   Find the cube contents
D To the G P   Above the side

 Ex.: An octagonal prism being 10 inches long, and 4 inches in the side, required the cube contents.

C   10 inches   772·5 cube inches content
D   * ·456 G P for 8 sides   4 inches side

PART III.

MECHANICS.

1st. THE MECHANICAL POWERS.

Levers are of three orders—

The First having the Fulcrum between the Weight and Power.

„ Second „ Weight between the Fulcrum and Power.

„ Third „ Power between the Fulcrum and Weight.

Let F represent the fulcrum or centre of motion, P the power, W the weight or resistance.

LEVER OR WHEEL AND AXLE.

General Formula for any order of Lever.

A Set the *distance* of P from F	Find W	or	Below W
B To the *distance* of W from F	Above P		Find P

Ex.: The arms of a lever are 36 and 3 inches respectively, a power of 40 lbs. acts on the longer arm; what weight will it balance?

A 36 in. dis. of P	W 480	also	Ratio 12
B 3 in. dis. of W	P 40		Above 1.

Note.—The velocity ratio, or the motion of power and weight, whether in questions of simple or compound machines, is always shown on A above unity on B.

COMBINATION OF LEVERS OR WHEELWORK.

General Formula.

A Product of dist. of powers } from F	Find W	and ratio
B „ „ weights }	Above P	above 1.

Ex.: The product of the distance of powers from F is 300 inches, and the product of distance of weights is 20 inches in a compound machine of this class, with a power of 80 lbs.; find weight and ratio of velocity.

A 15 ratio	First, Set { 300 in.	Find { W 1200 lbs.
B 1	20 „	P 80 „

PULLEYS AND COMBINATION.
General Formula.

A Set the number of pulleys   Find W
B To   1   Above P

INCLINED PLANE AND WEDGE.

A Set the length   Find W (or resistance)   and *r*
B To altitude or breadth   Above P   1

> Ex.: What power will support 1000 lbs. weight on an inclined
> plane 14 feet long, sloping 5 feet high?

A (2·8 *r*)   Set 14 ft.   W 1000 lbs.
B (1)   To   5 ,,   P 357.15 answer

THE SCREW.
General Formula.

A Set the circumference described by P   Find W
B To the distance (centre) of threads   Above P

> Ex.: The circumference described by a power equal to 6 lbs. is
> 36 inches, and the threads of a screw ½ inch from centres of
> edge; find the weight.

A Set 36 in. cirfce.   W 432 lbs.   *r* 72
B To ·5 in. thread   P   6 ,,   1

COMPOUND MACHINES.
General Formula.

A Set the product of ratios of the several parts   Find W   *r*
B To   1   Above P   1

GENERAL FORMULA FOR ALL MECHANICAL POWERS.

A Set the motion of power   ·   Find weight   and ratio
B To the motion of weight   Above power   1

Note.—Generally one-third more power than the results here found
(W and P being in equilibrium) is required to overcome resistance
owing to friction.

2ND. MACHINERY.

TOOTHED WHEELS (on A and B).

Note.—As the pitch of tooth, number of teeth, and diameter, are merely questions of simple proportion, and one of the terms, 3.14 = circumference of a circle whose diameter is 1, being constant, the operation in *all* cases is very easy, as the position of the terms in the statement are *fixed*, the conditions only varying.

General Rule.

A To 3.14 constant   then   Find N° of teeth   or   Below N° of teeth
B Set pitch of tooth   Above diam^r.   Find diam^r.

Or—

A Below 3.14   Set N°. of teeth
B Find pitch of tooth   To diam^r.

1. To find the number of teeth, pitch and diameter given.

Ex.: The pitch of tooth being 1.57 in., and diameter at pitch line 20 inches; required the number of teeth.

A To 3.14   40 No. of teeth answer
B Set 1.57   20 diam^r

2. To find diameter at pitch line, pitch and No. of teeth given.

Ex.: A pinion with 40 teeth, with 1.57 pitch, required the diameter at pitch line.

A To 3.14   40 teeth
B Set 1.57   20 diam^r. answer

3. To find the pitch of tooth, diameter at pitch line and N°. of teeth given.

Ex.: The diameter at the pitch line being 20 inches, and the N°. of teeth 40, required the pitch of tooth.

A Below 3.14   Set 40 teeth
B Find 1.57 pitch answer   To 20 in. diam^r.

REVOLUTIONS AND DIAMETERS:

1. To find the diameter of one wheel to any required number of
   tions in same time.

The revolutions of both wheels and } being known.
   the diameter of the other wheel }

(On A with slide *inverted*).

General Formula.

A   Find the ? reqd. diam^r.   Set ? the known diam^r.
S Inrd Above the ? reqd. revolt^{ns}.   To ? the known revolt^{ns}.

    Ex.: One wheel 180 inches diam^r., making 9 revolutions per
      minute, another is required to make with it 30 revolutions, re-
      quired its diameter.

A   54 required diam^r Set 18C
(S. In^r) 30 required revolutions   To   9

2. To find the diameters of two wheels to work at given velocities,

     The distance between the centres } being known.
     and number of required revolutions }

(Direct proportions on A and B).

General Formula.

A Find ? Radius   ? answer Set ? distance between centres
B Above ? *each* stated rev. ?   To ? the *sum* of revolutions

    Ex.: A shaft going 25 revolutions per minute, is to give motion
      to another of 15 revolutions in same time, the distance between
      centres 60 inches, required the diameters of the two wheels.

A Find 25 (required radii) and   42   Set 60 distance
B Above 15 (given rev.) and   25   To 40 *sum* of revolutions

3. To find the diameter required for a pulley to make more or less
   revolutions than one of stated diameter in equal time.

Revolutions and Diameter of one shaft } being known.
Revolutions only   of the other shaft }

(On A with Slide inverted).

General Formula.

A Set the diam^r known of one shaft   Find   ? the required diameter
B To the revoltn^s „   „   Above ? the given revolution
                                             of other shaft.

Example :

1 Shaft making 50 revolutions, with a drum 30 inches diameter.
1 Shaft   „   40   „

required the diameter for a drum on the second shaft, to drive the same *speed*, as 50 revolutions.

A   Set 30 inches dia.   Find   37.5 dia. for drum of 2nd shaft
(S. Int^r) To 50 revolutions   Above 40 revolutions

3RD. WEIGHT OF METALS.

THE full powers of the Slide Rule have never been fairly tested or employed for ready and instant calculations ; the ordinary formula is too much involved to make the operations apparent, as well as easy. It is only by availing ourselves of its proportional properties, that its capabilities stand in bold relief, either against brain or book ; while one or both may be challenged for speed, comprehensiveness, and accuracy ; and for simplicity the following method for finding the Weight of Metals cannot be approached. The proportions are evident, and the operation instant and unhesitating ; there is little to learn, and less to do, for the simple setting of the constants gives off the widest range of tabular results for each question of its class. All ordinary forms of the various metals in general use have been reduced to known proportions, which can be tested by mere inspection ; the amount of time and trouble saved is incredible to those who examine it.

The constants are formed for sectional areas, or diameters to one foot in length, and give the weight in lbs., cwts., or tons.

The sides of square sided metal and ⎱ are given in eighths, inches,
    the diameter of cylindrical metal ⎰   and feet,
so that the weight of the smallest rod, or of the largest mass or column, is arrived at with equal facility.

Use of the Constants explained.

Under the head of any metal, regard being had to its *form*, whether cubical or cylindrical, &c., will be found two constants showing the proportion of its area of section, or diameter, to weight in 1 foot lengths, which are to be set on the Rule, when the result is open to inspection; giving the weight of *all* sizes of that metal; the advantage is too evident to need urging.

Ex.: Required instantly, the weight per foot, of cast iron columns of *all* diameters.

Under *cylindrical* cast iron, inches in diameter, weight in lbs., we find

C Set 10. *meaning* that 10lbs. is the weight of

D To 2 *inches* diameter cylindrical iron.

Then with   Or, with weight in cwts. :

C Set 10   C is a line of lbs. weight .   C Set 2.2 cwts.

D To 2   D   „   diameters   D To 10 inches diam.

and as C D are in proportion of squares and roots, it follows that the weight of *all* diameters are obtained by *one* setting.

There is no variation from this simple rule for metals in lengths; and the weight of metals by surface is equally easy.

---

1. CUBICAL METALS, ALL 1 FOOT LENGTHS (on A and B).

Note.—Find the area of the end or section in eighths, inches, or feet.

Ex.: Side × side = area of section in square units.

Description of Metal.		Lead.	Copper and Gum Metal.	Cast Brass.	Wrought Iron.	Cast Iron.	Cast Zinc.	Water Compared.	Result.
Find weight *in lbs.* ....on A	Set	2	3	2	4	1	2	·6	lbs.
Above any section *in 8ths* ...on B	To	26	50	35	75	20	41	90	Sec.
Find weight *in lbs.* ....on A	Set	59	19	29	17	19	28	10	lbs.
Above any section *in in.* ...on B	To	12	5	8	5	6	9	23	Sec.
Find weight *in cwts.* ....on A	Set	4	2	1	3	2	1	1	Cwt.
Above any section *in in.* ...on B	To	91	59	31	100	71	36	260	Sec.
Find weight *in tons* ....on A	Set	7	10	7	5	10	10	2.5	Tons.
Above any section *in in.* ...on B	To	22	41	30	23	49	49	90	Sec.

* The specific gravity of *any other substance* being known, its weight can easily be found by comparison with the weight of water as given above, multiplying the weight of water thus obtained by the sp. gr. of the body of similar form and dimensions.

USEFUL DATA, showing Bases of the Constants, but not for use on the Rule.

For cubical metals.	Lead.	Copper and Gun Metal.	Brass.	Wrought Iron.	Cast Iron.	Cast Zinc.	Water.
Sp. gr. & oz. in cubic foot	11344	8788	8396	7788	7271	7190	1000
lbs. do.	709	549	524·75	486·75	454·44	449·37	62·5
cwts. do.	6·33	4·905	4·685	4·345	4·05	4·	·55
oz. in cubic inch	6·565	5·086	4·859	4·507	4·208	4·161	·579
lbs. do.	·4103	·3178	·3037	·282	·263	·26	·03617

2. CYLINDRICAL METALS, ALL 1 FOOT LENGTHS (on C and D).

	Lead.	Copper and Gun Metal.	Cast Brass.	Wrought Iron.	Cast Iron.	Cast Zinc.	Water.	Result.
Find weight *in lbs.* .... on C Set	6	3	4·5	1·5	1·9	3·8	·26	lbs.
Above any diam. *in 8ths* on D To	10	8	10	6	7	10	7	8ths.
Find weight *in lbs.* ... ... on C Set	35	12	26	24	10	22	8·5	lbs.
Above any diam. *in in.* on D To	3	2	3	3	2	3	5	in.
Find weight *in cwts.* ... on C Set	3·4	6	2·55	9·5	2·2	5	6	cwts.
Above any diam. *in in.* on D To	10	15	10	20	10	15	45	in.
Find weight *in tons* ...... on C Set	1	7	9	11	4	5·7	5	tons.
Above any diam. *in ft.* on D To	2	6	7	8	5	6	15	ft

Ex.: A cylindrical column of cast iron is 7 inches inside diameter, and 8½ inches outside; required its weight per foot in cwts.

C   Set 2.2   Find 1.08 cwt.   Find 1.6 cwt.
D   To 10   Above 7 inches.   Above 8.5 inches.

           1.6
           1.08 deduct hollow.
           ————————
           1.52 cwt. = 1.52 cwt. per foot.

USEFUL DATA, and Bases of the Constants for Cylindrical Metals.

	Lead.	Copper and Gun Metal	Cast Brass.	Wrought Iron.	Cast Iron.	Cast Zinc.	Water.
Ounces in a cylindrical foot of	8909·6	6902	6594	6117	5711	5647	755·4
lbs. " "	557·	431·2	412·	382·3	357·	353·	49·1
cwts. " "	4·972	3·852	3·68	3·412	3·18	3·14	·432
Ounces " inch	5·156	3·99	3·815	3·54	3·305	3·268	·4547
lbs. " "	·322	·2496	·2385	·2215	·2066	·204	·02842

3. SPHERICAL METALS, *any* diameter (on C and D).

		Lead.	Copper and Gun Metal	Cast Brass.	Wrought Iron.	Cast Iron.	Cast Zinc.	Water.	Result.
Find weight *in ounces* on C	Set	Diam.	Diam.	Diam.	Diam.	Diam.	Diam.	Diam.	oz.
Above same dia. *in 8ths* on D	To	12·2	13·9	14·1	14·6	15·2	15·3	41·2	8ths.
Find weight *in lbs.* ... on C	Set	Diam.	Diam.	Diam.	Diam.	Diam.	Diam.	Diam.	lbs.
Above same dia. *in in.* on D	To	2·16	2·45	2·5	2·6	2·7	2·72	23·	in.
Find weight *in cwts.* ... on C	Set	Diam.	Diam.	Diam.	Diam.	Diam.	Diam.	Diam.	cwts.
Above same dia. *in in.* on D	To	22·8	26·0	26 6	27·6	28·6	28·8	77·	in.
Find weight *in tons* ... on C	Set	Diam.	Diam.	Diam.	Diam.	Diam.	Diam.	Diam.	tons.
Above same dia. *in ft.* on D	To	2·45	2·82	2·86	3·0	3·08	3·1	83·	ft.

Ex.: A spherical ball of solid copper is 8 inches in diameter, required its weight in lbs.

C　Set 8 inches diameter.　　　Find 85 lbs. answer.
D　To 2.45.　　　　　　　　　Above 8 inches diameter.

Bases of Constants. Spherical metals.	Lead.	Copper and Gun Metal.	Cast Brass.	Wrought Iron.	Cast Iron.	Cast Zinc.	Water.
Ounces in a spherical foot	5940	4601	4396	4078	3807	3764	523·6
lbs. " "	371·2	287·5	274·75	254·9	239·	235·3	32·73
cwts. " "	3·314	2·57	2·453	2·275	2·131	2·1	·288
Ounces " inch	3·437	2·663	2·54	2·36	2 2	2·18	·3032
lbs. " "	·2148	·1664	·159	·1476	·1377	·136	·0189

BAR IRON, ALL 1 FOOT LENGTHS.

Note.—The weight of *all* metals *in lengths* could be found by the foregoing constants for cubical and cylindrical metal, but bar iron being an article of constant use, and requiring frequent reference, each kind is separately treated of.

1. FLAT BAR IRON.

One Foot long. From 1 sixteenth to 1 inch thick. Any ins. wide.

Sixteenths of inch thick.	$\frac{1}{16}$	$\frac{3}{16}$	$\frac{5}{16}$	$\frac{7}{16}$	$\frac{9}{16}$	$\frac{11}{16}$	$\frac{13}{16}$	$\frac{15}{16}$
Find lbs. wt. per ft. on A set	2·5	7·5	12·5	17·5	22·5	27·5	32·5	37·5
Above any ins. wide on B to	12	12	12	12	12	12	12	12

Eighths of inch thick	$\frac{1}{8}$	$\frac{1}{4}$	$\frac{3}{8}$	$\frac{1}{2}$	$\frac{5}{8}$	$\frac{3}{4}$	$\frac{7}{8}$ Inch.
Find lbs. wt. per ft. on A set	5	10	15	20	25	30	35	40
Above any ins. wide on B to	12	12	12	12	12	12	12	12

2. SQUARE BAR IRON.

Eighths Square.

C set 12   Find lbs. wt. per foot
D to 15   Above eighths sq. of sides

Inches Square.

C set 54   Find lbs. wt. per foot
D to 4   Above ins. sq.

3. ROUND BAR IRON.

Eighths in Diameter.

C set 4·2   Find lbs. wt. per foot
D to 10   Above any eighths diam.

Inches in Diameter.

C set 24   Find lbs. wt. per foot
D to 3   Above any ins. diam

TO FIND THE WEIGHT IN LBS. (AVDPSE.) OF ANY NUMBER OF SQUARE FEET SUPER. OF VARIOUS METALS, *one sixteenth thick,* FROM WHICH THE WEIGHT AT ANY THICKNESS MAY BE FOUND.

		Wrought Iron.	Cast Iron.	Cast Copper.	Gun Metal.	Cast Brass.
Find weight in lbs.	on A set	25	23	29	29	27 lbs.
Above any No. of sq. feet super.	on B to	10	10	10	10	10 sq. ft.

		Cast Lead.	Cast Zinc.	Cast Tin.	Cast Silver.	Pure Gold.
Find weight in lbs.	on A set	37	23	24	34	62·7 lbs.
Above any No. of sq. feet super.	on B to	10	10	10	10	10 sq. ft.

The easy method just given may fail to attract some who are wedded by habit to the use of the old gauge points for obtaining the weight of metals; but even these are greatly modified in the following Table, where, instead of taking the divisor on A, necessitating the use of four lines A B C D, the *square root* of it is employed. Every operation can then be performed on two lines (C and D) only, with equal accuracy and far greater readiness.

WEIGHT OF BODIES OBTAINED BY MEASUREMENT.

$$\text{Weight} = \frac{\text{Mean side or diameter}^2 \times \text{length.}}{\text{* Gauge Point.}}$$

* G P (divisor) on D is derived from the units of measure in an integer weight; viz., 2.16 is the square root of 4.66, the number of cubic feet of malleable iron in a ton. (See G P for iron, cubic, tons.)

Select the proper G P, considering whether for cube, cylinder, or sphere.
    ,,   dimensions in feet or inches.
    ,,   weight required in tons, cwts., or lbs.
Also for what kind of substance.

The Gauge Point having to be adapted to all these conditions, it is important to remember that the dimensions must be taken—
All in feet .... for tons.
Length in feet, and the ends or diameter in ins. for cwts.
All in inches........................... for lbs.

General Formula.

C Set the length   Find the weight on C.

D To the proper G P   Above the m. square or diameter

The *mean square* of cubical forms must be used.
The mean diameter of cylindrical forms must be used.

Table for Weight of Bodies by Measurement.

	CUBIC.			CYLINDER.			SPHERE.
	Tons.	Cwts.	lbs.	Tons.	Cwts.	lbs.	lbs.
Lead . . .	1·777	4·768	1·562	2·	5·38	1·76	2·159
Gun Metal .	2·01	5·39	1·77	2·268	6·1	1·99	2·44
Copper, cast .	2·02	5·42	1·78	2·28	6·12	1·9	2·45
Brass, ,, .	2·14	5·74	1·88	2·41	6·48	2·12	2·6
Iron, mal. .	2·16	5·8	1·9	2·44	6·55	2·21	2·66
Iron, cast . .	2·232	5·99	1·96	2·52	6·75	2 22	2·72
Zinc, ,,	2·236	6·	1·96	2·523	6·77	2·21	2·71
Marble .	3·601	9·66	3·19	4·1	10·9	3·6	4·4
Portland .	3·742	10·	3·27	4·22	11·33	3·7	4·5
Chalk . .	3·571	9·58	3·45	4·03	10·8	3·89	4·77
Clay . .	4·233	11·36	3·72	4·78	13·35	4·2	5·14
Gravel . .	4·32	11·59	3·8	4·87	13·1	4·28	5·25
Sand . .	4·32	11·59	3·8	4·87	13·1	4·28	5·25
Bricks . .	4·123	11·06	3·85	4·65	12·48	4·34	5·32
Earth . .	4·54	12·2	3·96	5·12	13·74	4·47	5·48
Coal . .	5· 4	14·33	4·69	6·02	16·16	5·29	6·48
Water . . .	5·98	16·05	5·253	6·75	18·1	5·93	7·26
Oak seasoned	6·93	18·58	5·46	7·8	20·96	6·16	7·55
Ash . . .	6·52	17·49	5·76	7·36	19·74	6·5	7·96
Elm . . .	8·12	19·59	6·4	8·24	22·1	7·4	8·86
Memel Fir .	8·22	21·76	7·13	9·27	24·5	8·	9·85
Yellow Pine .	8·59	22·5	7·85	9·7	25·5	8·85	10·84
The dimensions must be given.	All in feet for tons.	Feet long inches side for cwts.	All in inches for lbs.	All in feet for tons.	Feet long inches side cwts.	All in inches for lbs.	All in inches for lbs.

(Marble through Coal bracketed as "Average")

Example of Working.

A cylindrical hollow column of cast iron is 15 feet long; inside diameter, 9 inches; outside diameter, 10 inches; required its weight in cwts.

For the right G P. look under cylinder, under cwts. for cast iron, find 6.75.

C Set 15. feet long   26.62   32.87 cwt.
D To   6.75 G P.   9 in.   10 in.

[Answer on next page.

Note. — When *any* length
is once set to a G P, the weight
to *all* diameters for that length
is given.

32.87
Deduct hollow 26.62
————
Weight of column 6.25 cwt.

Data to form Gauge Points, if required, for the following Metals :—

	Specific Gravity.	Lbs. in a Cubic ft.	Oz. in a Cubic inch.
Platinum, pure	21500	1343·8	12·44
Gold, do.	19258	1203·6	11·14
Gold, hammered........	19361	1210·7	11·26
Silver, pure	10474	654·6	6·06
Silver, hammered	10510	656·9	6·08
Mercury, fluid	13568	848	7·85
Tin, cast	7292	455·7	4·22
Aluminum ...... average	2600	162·5	1·5

To FIND THE WEIGHT IN LBS. OF ANY NUMBER OF SQUARE FEET SUPER. OF SHEET IRON, THE THICKNESS AGREEING WITH THE NUMBER ON THE WIRE GAUGE.

		Numbers on Wire Gauge.						
Thickness Wire Gauge		1 $\frac{1}{18}$	2	3	4 $\frac{1}{8}$	5	6	7
Find weight in lbs. .... on A Set		125	120	110	100	90	80	75
Above any No. of square feet on B To		10	10	10	10	10	10	10

		Numbers on Wire Gauge.						
Thickness Wire Gauge		8	9	10	11 $\frac{1}{8}$	12	13	14
Find weight in lbs. .... on A Set		70	60	56·8	50	46·2	43·1	40
Above any No. of square feet on B To		10	10	10	10	10	10	10

		Numbers on Wire Gauge.						
Thickness Wire Gauge		15	16	17	18	19	20	21	22
Find weight in lbs. .... on A Set		39·5	30	25	21·8	19·3	16·2	15	13·7
Above any No. of square feet on B To		10	10	10	10	10	10	10	10

Ex.: Required the weight in lbs. of 27 feet super. of No. 10 gauge sheet iron.

A   To 56.8 lbs.   Find   153.4 lbs.   (Weight in lbs.)
B   Set 10 sq. feet.   Above   27 sq. feet.   (or *any* super. feet.)

To find the Weight in Lbs. of Lead, Copper, and Wrought Iron Pipes, 1 Foot Long, and any Circumference in Inches, 32nds Thick.

								32nds.
Thickness in 32nds of an inch					1	2	3	4
Lead	Find weight in lbs.	on	A	Set	2	4	6	8
	Above circumference in inches	on	B	To	13	13	13	13
Copper	Find weight in lbs.	on	A	Set	1·7	3·4	4	3·4
	Above circumference in inches	on	B	To	14	14	11	7
Wrought Iron	Find weight in lbs.	on	A	Set	2·5	2·5	5	5
	Above circumference in inches	on	B	To	24	12	16	12
Thickness in 32nds of an inch					5	6	7	8 ¼ in.
Lead	Find weight in lbs.	on	A	Set	10	12	14	16
	Above circumference in inches	on	B	To	13	13	13	13
Copper	Find weight in lbs.	on	A	Set	8·5	8	11	29
	Above circumference in inches	on	B	To	14	11	13	30
Wrought Iron	Find weight in lbs.	on	A	Set	14	10	8	10
	Above circumference in inches	on	B	To	26	16	11	12

Ex.: Required the weight in lbs. of 1 foot of copper pipe, 6 32nds thick, 20 inches circumference.

A	To	8 (lbs. per foot long)	Find	14.5 lbs.
B	Set	11 (in. circumference.)	Above 20 in. circumference.	

To find the Weight in Lbs. of Cast Iron Pipes, 1 Foot Long, any Circumference in Inches, from ⅛th to 1¼ Inch Thick.

Thickness	⅛	¼	⅜	½	⅝	¾	⅞	Inch.	1⅛	1¼
Find weight in lbs. .... on A Set	7	7	20	16	22	28	33	32	36	40
Above the mean circumfe. in ins. on B To	18	9	17	10	12	12	12	10	10	10

Note.—Two flanges generally reckoned as 1 foot.

Ex.: Required the weight in lbs. of 1 foot long of cast iron pipe, ⅝ths thick, 30 inches mean circumference.

A	To	22	Constants	Find 55 lbs.
B	Set	12	for ⅝ thick	Above 30 mean circumference

Armour Plating, Weight in Lbs. and Cwts., of any Number of Super. Square Feet, from 1 to 10 Inches Thick (on A and B.)

Result in Lbs.

Inches thick				1	1½	2	2½	3	3½	4
Find weight in lbs.	on	A	Set	406	608	812	1014	1217	1420	1622
Above *any* sq. ft. super.	on	B	To	10	10	10	10	10	10	10
Inches thick				4½	5	5½	6	6½	7	7½
Find weight in lbs.	on	A	Set	1825	2028	2230	2434	2637	2840	3042
Above *any* sq ft. super.	on	B	To	10	10	10	10	10	10	10
Inches thick				8	8½	9	9½	10	12	
Find weight in lbs.	on	A	Set	3245	3448	3650	3854	4056	4868	
Above *any* sq. ft. super.	on	B	To	10	10	10	10	10	10	

Ex.: Find weight in lbs. of armour plate, 4½ inches thick, 9 feet by 3 = 27 square feet.

A	Set	1825 lbs.	Find	4927 lbs.	
B	To	10	Above	27	square feet.

Result in Cwts.

Inches thick...............				1	1½	2	2½	3	3½	4
Find weight in cwts. ...	on	A	Set	3·62	5·43	7·25	9·05	10·86	12·67	14·5
Above *any* sq: ft. super.	on	B	To	10	10	10	10	10	10	10
Inches thick				4½	5	5½	6	6½	7	7½
Find weight in cwts. ...	on	A	Set	16·3	18·1	19·9	21·7	23·5	25·34	27·16
Above *any* sq. ft. super.	on	B	To	10	10	10	10	10	10	10
Inches thick...............				8	8½	9	9½	10	12	
Find weight in cwts. ...	on	A	Set	28·97	30·78	32·58	34·39	36·2	43·4	
Above *any* sq. ft. super.	on	B	To	10	10	10	10	10	10	

Ex.: Find the weight in cwts. of armour plate, 4½ inches thick, $9 \times 3 = 27$ square feet.

A	Set	16.3 (cwt.)	Find	44 cwt.
B	To	10 feet square.	Above	27 super. square feet.

With *weight only given*, in Lbs. of Spherical Cast Iron, to find Diameter.

A To 7·55	Find a product * the cube root of which = dia.
B Set 1.	Above the lbs. weight given

* Find the cube root of this product by cube root on $\frac{C}{D}$

Ex.: Find the diameter of a cast iron ball 28.5 lbs. weight.

A To 7.55	Find 216 $\sqrt[3]{\ }= 6$ inches diameter of ball.
B Set 1	Above 28.5 lbs. given.

STEAM ENGINES, &c.

THE ordinary rules are for nominal horse power only, and do not give a fair approximation even to the effective work of an engine. The following useful Slide Rule formula is adapted from the arithmetical process given in Templeton.

CONDENSING ENGINES (on C and D).

First. To the pressure in lbs. per *circular inch* × ·8 = ?

Add 3.73.

Sum.

Multiply by the speed in feet per minute ?

Product* ?

C Set the product *   Find the effective horse power
D To   182   Above (any) diam. of cylinder in inches.

Ex.: The pressure of steam being 6 lbs. on the *circular* inch, the speed of the piston 180 feet per minute, and the diameter of the steam cylinder 40 inches, find the effective horse power.

lbs. 6 × .8 = 4.8, add 3.73 = 8.53

Speed   180 ×

Product* 1535

C Set * 1535   74.4 effective horse power, answer.
D   182   40 in. diameter.

D

HIGH PRESSURE ENGINES (on C and D).

First. From the pressure in lbs. per circular inch   ?
     Deduct due to the work of the engine itself   4 lbs.

                       Effective steam   Remainder.
Multiply by the speed of the piston in ft. per minute   ?

                                         Product.

C   * product.   Find the effective horse power
D   182   Above (any) diam. of cylinder in inches.

, *Ex.:* The pressure of steam being 30 lbs. on the circular inch, the speed of the piston 180 feet per minute, and the diameter of the steam cylinder 20 inches, find the effective horse power.

                lbs. 30, less $4 = 26 \times 180 = 4680$ product

C   Set   4680   57 horse power, answer
D   To   182   20 inches diameter.

To find pressure in lbs. per inch, Or diam. of cyl. to cwts. lift req^d.

(S. Invt^d) Set cwts. lift.   Find effective press. lbs. per inch
D   To 14.2   Above diam. of cylinder in inches.

Surface of water in boilers required to supply steam to cylinders of any diam. in inches.

C   Set   13   Find square feet of water surface required in boiler
D   To   10   Above diam. of any steam cylinder in inches.

Pressure of steam on square and circular inches compared.

A   Set   70   Pressure of steam on square inches }
B   To   55   Ditto   circular inches } compared.

A   Set   45   Find column of mercury supported inches
B   To   22   Above pressure of steam, lbs. per *square* inch.

A   Set   53   Find column of mercury supported inches
B   To   20   Above pressure of steam, lbs. per *circular* inch.

REGULATORS (on A and B).

When the velocity (revolutions per minute) of the governor is given, to find length of pendulums.

A   Find * ? == square root of length.   (To   187.5
B   Above 1   1st. { Set the number of
                                         (revolutions per minute.

* For the square of which (the length of pendulums required) look to C and D. •

Ex. : The velocity of a governor being 40 revolutions per minute, find the length.

A   Find 4.69 squared 22 in. length.   To   187.5
B   1   Set 40 revolutions.

When the length of the pendulums is given, to find the revolutions.

A   Find No. of revolutions.   To 187.5
B   Above 1.   Set ? square root of given length.

Ex. : The length of the pendulums being 22 inches, find the revolutions.

A   Find 40 revolutions, answer.   187·5
B   1   4·69 sq. root of 22, found
                                                     on C and D.

CONSUMPTION OF FUEL, BY TONS OR CWTS., FOR MILES OR HOURS RUN.

Answer in lbs. per mile or hour.

SUPPLY IN CWTS.

A   Below 1.12.   To the number of miles or hours
B   Find lbs. per mile   Set the cwts. consumed.

SUPPLY IN TONS.

A   Below 22.4.   To the number of miles or hours
B   Find lbs. per mile   Set the tons consumed.

PART IV.

APPLICATION OF THE SLIDE RULE TO BUILDERS' WORK; FOR CIVIL ENGINEERS, SURVEYORS, CONTRACTORS, &c.

REDUCTION OF BRICKWORK FROM SUPER. OF VARIOUS THICKNESSES TO STANDARD (1½) CUBE FEET, CUBE YARDS, &c. (on A and B.)

Brickwork of various thicknesses reduced to Standard and Cube Work, &c.

No. of bricks thick	½	1	1¼	1½ (Standard)	1¾	2	2¼	2½	2¾	3
Compare										
Feet super....... on A Set	300	300	60	100	60	75	60	60	54	50
Feet standard... on B To	100	200	50	100	70	100	90	100	100	100
Feet super....... on A Set	132	66	53	80	38	33	29	45	24	22
Feet cube ...... on B To	50	50	50	90	50	50	50	35	50	50
Feet super. ...... on A Set	240	180	140	120	28	90	80	70	65	60
Yards cube ... .. on B To	5	5	5	5	4	5	5	5	5	5
Feet super. ...... on A Set	816	408	326	272	234	204	182	163	148	136
Rods standard... on B To	1	1	1	1	1	1	1	1	1	1
Feet super. ...... on A Set	20	20	40	20	26	20	20	20	20	20
No. of bricks...... on B To	110	220	550	330	500	440	500	550	600	661

Work given in inches thick	4½	5	9	9½	10	11	12	13	14
Feet super.............. on A Set	50	42	70	37	35	45	35	65	50
Feet standard............ on B To	16	15	45	25	25	35	30	60	50
Feet super.............. on A Set	40	60	80	38	60	55	50	55	60
Feet cube .... on B To	15	25	60	30	50	50	50	60	70

Work given in inches thick	14½	15	16	17	18	18½	19	20
Feet super.............. on A Set	39	30	35	33	35	30	37	35
Feet standard......... .. on B To	40	32	40	40	45	40	50	50
Feet super.............. on A Set	75	60	60	60	60	65	60	60
Feet cube .... on B To	90	75	80	85	90	100	95	100

When any of the above proportions either for bricks or inches thick are once set, the lines $\frac{A}{B}$ form Tables for comparing and reducing all super. of that thickness.

Note.—As reference may not be made to other parts of the work unconnected with building operations, it is re-stated that nothing is required but to set the constants found under any given thickness and opposite the reduction required, viz., to standard, cube, &c.

General Formula.

A	Set	? The constants		A is then a line feet super	
B		? For thickness		B „ reduced work	

Ex. : Reduce 30, 80, 120, 250 feet super. 2½ bricks thick,

1st, to 1½ bricks (standard).

A Below	30	80	120	250 of 2½ bricks	Set 60
B Find	50	133.3	200	416.6 of 1½ bricks	To 100

2nd, to cube feet

A Below	30	80	120	250 of 2½ bricks	Set 45
B Find	56.6	151.1.	226.6	472.2 of cube feet	To 85

Decimal values of any feet in a rod.

A Find the decimal parts	Below any decimal parts	1 To	
or			1st
B Above any No. of feet	Find the value in feet	272 Set	

TABLES OF USEFUL PROPORTIONS FOR BUILDERS.

A	Set	8̄	Below £ price per rod 1½ work
B	To	7	Find pence price per foot super.

A	Set	23	Below shillings per cube yard
B	To	13	Find £ cost per rod.

A	Set	50	Below shillings per load of 50 feet
B	To	12	Find pence per cube foot.

A   Set   60   Below cubic yards
B   To   23   Find 1000ˢ of bricks required.

A   Set   34   Below cubic yards
B   To   3   Find rods standard (1½).

A   Set   90   Below cubic feet
B   To   8̄0̄   Find standard feet (1½).

A   Set   34   Below cube yards
B   To   39   Find tons of brickwork.

A   Set   70   Below cube feet
B   To   60   Find cwts. of brickwork.

A   Set   45   Below number of bricks (5lb.)
B   To   2   Find cwts.

A   Set   9   Below feet super.
B   To   36   Find No. of paving bricks flat 9 × 4 in.

A   Set   9   Below feet super.
B   To   48   Find No. of paving bricks on edge 9 × 3 in.

A   Set ·   8   Below £s cost per acre
B   To   12   Find pence cost per square rod.

A   Set 123   Below £s cost per acre
B   To   6   Find pence cost per square yard.

A   Set   30   Below shillings cost per square rod
B   To   9   Find pence cost per square yard.

A   Set   50   Below shillings cost per load 50 feet
B   To   12   Find pence cost per cube foot.

TIMBER MEASURING.

(For superficial measurement, see Mensuration of Areas.)

1. Round Timber.

Note.—Instead of finding the quarter girth, it is easier by the Slide Rule to use a gauge point answering to the *whole girth*, *i.e.*, the G P for quarter girth, 12 × 4 = 48, the G P for whole girth. Both methods are shown.

The customary measure gives only about four-fifths of the *true* contents of round timber, proportioning it to timber when hewn; gauge points for the true contents are given also.

Allowance for Bark.	Customary Measure.		True Contents.
	Gauge Points for Whole Girth.	Gauge Points for Quarter Girth.	Gauge Points for Whole Girth.
None	48 on D	12 on D	42.5 on D
1-eighth	54.8	13.7	48.6
1-tenth	53.34	13.33	47.23
1-twelfth	52.36	13.1	46.37

Ex.: A felled tree is 40 feet long, its mean girth 80 inches, no allowance for bark.

C Set 40 feet length   Find 111·1 cubic feet.
D To 48 G P.*   Above 80 ins. *whole* girth.

Common Method.

C Set 40 feet length   Find 111·1 cubic feet.
D To 12   Above 20 ins. *quarter* girth.

True Contents.

C Set 40 feet length   Find 142 cubic feet.
D To 42·5   Above 80 ins. whole girth.

2. Hewn Timber.

The usual method of measuring hewn timber is given below; but being satisfied that the Slide Rule is never more advantageously employed, nor easier learned, than when the *proportions* of a question are shown in the working, I recommend the following form, which answers for timber of equal or unequal sides, and for any size.

A To 144   Below *any* square inches in section
B Set length in feet.   Find cube feet contents.

For this purpose, and other questions in cubed timber, the 144 is marked near the centre of A, which gives the whole range of the line, the reading of the numbers on A and B then being

A 10 20 to 100   (144) 200 &c. to 1000 For sections on A.
B 1 2   10   20   . 100   For length and cube
                                          contents on B.

The simplicity of the process is apparent, for when 144 square inches in the cross section of any sided timber gives 1 foot cube for each foot in length, any given length in feet being set to 144, a table of sections on A and cube contents on B is formed; every increase to the right and decrease to the left of 144, marking its own proportion in cube contents. For an easy example take three sections each to 30 feet in length, viz.: 9 × 8 = 72, 12 × 12 = 144, and 12 × 24 = 288 inches.

A Below 72   To 144   Below 288
B Find 15 cubic feet Set 30 feet length. Find 60 cubic feet

Find 30 feet cube.

And so of any section, to any length set to 144, it being only necessary (before setting the length) to square up the inches in sides by common multiplication on A and B, as 20 × 25 = 500 sq. ins. in the cross section.

A To 20   Find 500 sq. inches in section
B Set 1   Above 25.

Note.—Smaller wood is sorted and sized for fuel into—
  Shids   4 feet long and 16, 23, 28, 33, and 38 inches girth
  Billets   3   „   7, 10, 14   „
  Cord wood, stacked, 3 × 3 × 14.22   128 c. ft. = 1 cord.
    Ditto   measuring 4 × 4 ×   8 = 128   = „
  Billets   100   = „
  Ditto,   cwt. 10   = „
  Bavins and spray   100   = 1 load.

COMMON METHOD FOR HEWN TIMBER.

C Set the length in feet   Find cube feet content
D To 12.   Above the *mean sq.* in inches.

To find the mean square, say of 20 × 30 inches.

C Set 20 the smaller side   Below 30 the greater side
D To 20 the like number.   Find 24.5 the mean square.

Note.—In hewn timber of unequal sides, by the common method, the mean square must be taken. For example, with a section 30 in. by 20 in., the *true* mean square is 24.5, and *not* 25, as is sometimes erroneously used. The result in cubic measurements is very different, for in 40 ft. length of this section the

True mean square, 24.5, gives 166.6 content,

Arithmetical mean of 25 gives 173.2 content,

and the greater the difference of the sides the greater the error.

SCANTLINGS.

The rule given for Hewn Timber answers for all Scantlings,
or

A is a table of feet run   To 144 (marked on *centre*)
B is a table of feet cube   Set the square inches in sections.

Ex.: Find the cube contents to feet run of scantling, 6 × 6 inches = 36 ins. section. This easy example is given, but all others follow the same rule.

A Below 20 feet run 30 40   To 144
B Find   5 ft. cube 7.5 10 $^{\&c.}$ Set 36 the sq. ins. in cross sec.

To find the price per foot run of any scantling in the proportion to any price per foot cube.

A Below any sq. ins. in section   To 144
B Find pence per foot run.   Set price in *pence* per foot cube.

Ex.: Timber at 3*s.* 6*d.* = 42d. per foot cube, find the price per foot run of scantling, 2½ × 4 = 10, 3 × 4 = 12, and 3 × 5 = 15.

A Below 10   12   15   ins. section   ⎧To 144
B Find 2.9   3.5 4.4 pence per ft. run. 1st.⎨Set 42 pence per foot
                                                      ⎩   cube.

REDUCTION OF DEALS AND BOARDS TO ST. PETERSBURG
STANDARD, 120 deals of 12′ 11″ × 1½″.

Feet run of any section reduced to feet run St. Petersburg.

Section in inches				4×12	4×11	4×10	4×9	4×8
Find run in feet of any section...... on	A	Set		495	540	594	660	742·5
Above run in feet St. Petersburg ... on	B	To		1440	1440	1440	1440	1440
Section in inches				4×7	4×6	3×12	3×11	3×10
Find run in feet of any section...... on	A	Set		848	990	660	720	792
Above run in feet St. Petersburg ... on	B	To		1440	1440	1440	1440	1440
Section in inches				3×9	3×8	3×7		3×6
Find run in feet of any section...... on	A	Set		880	990	1131·4		1320
Above run in feet St. Petersburg ... on	B	To		1440	1440	1440		1440
Section in inches				$2\frac{1}{2}\times12$	$2\frac{1}{2}\times11$	$2\frac{1}{2}\times10$		$2\frac{1}{2}\times9$
Find run in feet of any section...... on	A	Set		792	864	950·4		1056
Above run in feet St. Petersburg ... on	B	To		1440	1440	1440		1440
Section in inches				$2\frac{1}{2}\times8$	$2\frac{1}{2}\times7$	$2\frac{1}{2}\times6\frac{1}{2}$		$2\frac{1}{2}\times6$
Find run in feet of any section...... on	A	Set		1118	1357·7	1462		1584
Above run in feet St. Petersburg ... on	B	To		1440	1440	1440		1440

Example: Reduce 2000 feet run of 3×9 to feet run of St. Petersburg.

A   Set   880   Below 2000 or *any* feet run of 3×9
B   To   1440   Find 3300 feet St. Petersburg.

Feet Run of any Section reduced to St. Petersburg Deals.

Section in inches				4×12	4×11	4×10	4×9	4×8
Below feet run of any section...... on	A	Set		165	90	495	110	310
Find No. of St. Petersburg deals on	B	To		40	20	100	20	50
Section in inches				4×7	4×6	3×12	3×11	3×10
Below feet run of any section...... on	A	Set		460	750	110	120	330
Find No. of St. Petersburg deals on	B	To		65	91	20	20	50
Section in inches				3×9	3×8	3×7		3×6
Below feet run of any section............ on	A	Set		220	330	330		330
Find No. of St. Petersburg deals ...... on	B	To		30	40	35		30

Feet Run of any Section reduced to St. Petersburg Deals.

			2½×12	2½×11	2½×10	2½×9
Section in inches			2½×12	2½×11	2½×10	2½×9
Below feet run of any section............ on	A	Set	330	360	230	260
Find No. of St. Petersburg deals ...... on	B	To	50	50	29	30

			2½×8	2½×7	2½×6½	2½×6
Section in inches			2½×8	2½×7	2½×6½	2½×6
Below feet run of any section............ on	A	Set	495	430	390	660
Find No. of St. Petersburg deals ...... on	B	To	50	38	32	50

Ex.: How many St. Petersburg deals are there in 550 feet run of 4 × 9.

A Set 110   Below 550
B To 20   Find 100 St. Petersburg deals.

This table is limited to a few of the most useful sizes; but by a little ingenuity *any conceivable section* can be *instantly* reduced to St. Petersburg standard, by the following

General Rule.

Note.—Although four lines are employed, C and D are only *once set*, as a register. D line being only used to set 5.1, the constant to 10, which cannot be reached on A. Note how the operations follow, 1st, 2nd, and 3rd.

A 2nd. { To *any* cross section (in. × in.) on A.
B   { Set 10 (mid. No.) on B.   3rd. { Below any No. of ft. run on B.
C   { Find No. of St. Ptbg Dls. on C.   1st. { Set 5.1 on C.
D   { To 10 on D.

Ex.: Reduce 330 feet run of 3×10 to St. Petersburg standard. Find how many deals (worked as above).

A 2nd. { To 3×10 in. = 30 sq. in. in section.
B   { Set   10   3rd. { 330 feet run.
C   { 50 deals, answer.   1st. {   5.1
D   {   10

Note.—The St. Petersburg standard consists of 120 deals 12 feet long 11' × 1½ inches.

Section	=	16.5	Square inches.
Total length	=	1440.	Feet run.
Ditto	=	17280.	Inches run.
Solidity	=	285120.	Inches cube.
Ditto	=	165.	Feet cube.
1 Deal	=	1.375	,, ,,
1 Deal	=	2376	Inches cube.

PRICES OF DEALS, &c., of *all sizes* (120 of 12 feet), compared with St. Petersburg standard at *any given rate.*

Below *any other section.*

A 2×6=12   1st. Set   16.5   as 3×8 = 24   3×12 = 36
B   £14·54   {To (say £20 } Find   29·1   £43·63.
       £14 10s. 11d. {Or *any* other rate.} Equal to £29 2s.   £43 12s. 6d.

To find the price per load of 50 cube feet, agreeing with St. Petersburg standard, or 120 of 12 feet, at any given rate.

A   Below *any* given rate   1st. Set   32
B   Find £ per load of 50 cubic feet   To   9.7 = £9 14s.

To find instantly the number of feet run of quartering of *any* section, to a load of 50 cubic feet.

A   Find feet run to a load   To 144.   Find feet run to a load.
B   Above 50 on B.   1st. {Set ? in. × in. } Above 50 on B.
                                      {Of any section. }

Note.—It is simple but useful to observe that in 120 of 12 feet of any size the cube feet contents equal 10 times the section.

Ex. : 3 in. × 12 in. = 36 = 360 cubic feet.   4 × 11 = 44 = 440 cubic feet.

AVERAGE DIMENSIONS FOR SCANTLINGS OF ORDINARY STRENGTH (on C and D.)

	Gauge Points for proper Section to Length.		
	Deal.	Elm.	Oak.
Ceiling joists	245	220	194
Common joists	380	342	300
Rafter	333	300	265
Do. principal	600	540	475
Common beams	512	460	410
Purlin	725	650	575
Summer	1080	970	860

General Formula.

C   Set *any* inches thick   Below the length in feet
D   To G P.   Find inches depth required.

AVERAGE STRENGTH OF ROPES, CHAINS, RODS, ETC.
(ON C AND D).

————

ROPES.

C	Set 6	Find weight in lbs. per fathom.
D	To 5	Above circumference in *inches*.
C	Set 7	Find cwts. safe bearing.
D	To 2	Above circumference in *inches*.
C	Set 5	Find tons safe bearing.
D	To 7	Above circumference in *inches*.

CHAINS.

C	Set 2	Find the lbs. weights per fathom.
D	To 3	Above diameter in *sixteenths*.
C	Set 12	Find cwts. safe bearing.
D	To 4	Above diameter in *eighths*.

To find breaking weight in tons:—

Linked.			*Studded.*	
C Set 4 Find tons br. weight		C Set 8 Find tons br. weight.		
D To 6 Above sixteenths dia.		D To 8 Above 16″ diam.		

To find proof strength :—

Linked.			*Studded.*	
C Set 5 Find tons proof		C Set 7 Find tons proof		
D To 10 Above 16″ diam.		D To 10 Above 16″ diam.		

IRON RODS.

Square		*Round*	
C Set 20 Find cwts. safe bearing		C Set 15 Find cwts safe bearing	
D To 3 Above *side* in eighths		D To 3 Above *diam.* in eighths.	

BEAMS (1 *inch thick*).

Gauge points for Iron.	Oak.	Pine.	Fir.
To be used on C. 7.6	1.25	1	.75

GENERAL FORMULA FOR BEAMS SUPPORTED BOTH ENDS.

First find square root of length given; C and D gives this by *inspection ;* then

C Set G P for kind	Find cwts. safe load for beams 1 *inch thick.*
D To *square root* of feet length	Above inches depth of beam.

Ex.: What is a safe load for an iron beam 2 inches thick, 6 inches deep (16 feet long, square root 4).

C Set 7.6 G P iron	17 cwt. for *each inch* thickness
D To 4, square root of length	6 inches deep.

CONTENTS of cylinders at 1 *foot deep,* diameter *in feet.*

C Set 42	Find cube yards at 1 foot deep
D To 38	Above any diam. of cylinder in feet.
C Set 7	Find cube feet at 1 foot deep
D To 3	Above any diameter in feet.
C Set 490	Find gallons at 1 foot deep
D To 10	Above any diameter in feet.

CONTRACTORS' WAGES TABLE (on A and B,)

Showing the amount in Shillings and Pence for any number of hours employed, at from 1 to 12 pence per hour.

Rate per hour in pence...			1	1½	2	2½	3	3¼	4	4½
Find shillings' amount... on A	Set		2	3	4	5	6	7	8	9
Above No. of hours' work on B	To		24	24	24	24	24	24	24	24
Rate per hour in pence...			5	5½	6	6½	7	7½	8	8½
Find shillings' amount ... on A	Set		10	11	12	13	14	15	16	17
Above No. of hours' work on B	To		24	24	24	24	24	24	24	24
Rate per hour in pence...			9	9½	10	10½	11	11½	12	
Find shillings' amount ... on A	Set		18	19	20	21	22	23	24	
Above No. of hours' work on B	To		24	24	24	24	24	24	24	

Ex.: Wages due 40 hours' work at 4½*d*.

    A Set 9 Find 15*s*.   Find shillings due
    B To 24 Above 40, and above *any* hours at 4½*d*.

It is evident that a large pay sheet at various rates and hours may be made up and checked in a few minutes with the aid of the Slide Rule.

MANUAL AND HORSE LABOUR (on A and B.)

Gauge Points or Constants for Manual Labour.	Work Required.	Gauge Points or Constants for Horses' Work.
4000	Cubic feet of water raised	25300
148	Cubic yards ,, ,,	935
2232	Cwts. weight ,,	14100
112	Tons weight ,,	705
74	Cubic yards of earth ,,	475

1 foot high per hour.

General Formula.

A To G P   Work due
B Set feet high to be raised   Hours' labour.

Ex.: How many cubic yards of earth, to be raised 16 feet, are due to 20 hours' manual labour.

A G P 74   92 cube yards, answer
B   16 feet   20 hours.

Ex.: How many hours' manual labour are required to raise 92 cube yards of earth 16 feet high?

A G P 74   92 cube yards
B   16 feet   20 hours, answer.

PART V.

COMMERCIAL ARITHMETIC.

EXCHANGE OF MONEY, AND CONVERSION OF ENGLISH AND FOREIGN WEIGHTS AND MEASURES, WITH ENGLISH AND FRENCH STANDARDS IN DETAIL.

Note.—The proportional properties of the Slide Rule render invaluable assistance in Commercial Arithmetic, the most opposite denominations being by it as easily convertible as the more simple ones, altogether obviating the tedious process of reduction, as well as the necessity for bulky tables. Through the aid of fixed proportions, *all* others *of that kind* are instantly constructed by merely setting the rule as directed; by this simple method *any* Foreign and English standards may be compared as easily as if all were printed in a tabular form, an impossibility with variable exchanges, while, with the Slide Rule, *any* rate may be assumed at pleasure. The Examples given are arranged to convey a certain amount of information as to the relative value. Two or three points only demand attention.

1st. The line A always represents Foreign money, weight, &c.
    „   B always represents English   „

2nd. The proportions given must be maintained throughout the scale in each operation.

Example, showing the use of the constants and the simplicity of conversion. See French Measures, page 69.

A Set 76 Line of centimetres   100   120   450 cent^m
                                        thus,
B To 30   „   inches   39.3   47.2   177 inches

Any constants, with the denominations affixed, are meant to show that the whole lines, right and left, are to be read as numbers of that kind. *Example :*

A 3   9   Set 12 gulden   30   36 gulden
                1st ||
B 5   15   To 20 shillings   50   60 shillings

Nothing is required but to set the numbers and observe the *proportions* throughout; all the commercial tables that follow are formed on this principle, which has been dwelt on to prevent the necessity for further explanation.

Country.	Money of Account. Gold & Silver.	Average Current Value. s. d.	Comparison with the £ English. (*Any* Rate of Exchange may be set to the £ sterling in lieu of the following.)
AMERICA, U.S.	1 dollar 100 cents.	4 2	A Set 4.80 dollars B To 1 £ English
AUSTRIA	1 gulden 100 nenkreuzers	2 0	A Set 10 gulden B To 1 £
BELGIUM	1 frank 100 cents.	0 9½	A Set 25 francs B To 1 £
BRITISH INDIA	1 rupee 16 annas	2 0	A Set 10 rupees B To 1 £
DENMARK	1 rigsdaler 96 skilling	2 3	A Set 9 B To 1 £
FRANCE	1 franc 100 centimes	0 9½	A Set 25 francs B To 1 £
GERMANY	1 gulden 60 kreutzers	1 8¼	A Set 11.50 gul. & kr. B To 1 £
HOLLAND	1 gulden 100 cents.	1 8¼	A Set 11.85 gulden B To 1 £
ITALY	1 lira 100 centissimi	0 9½	A Set 25 lira B To 1 £
PORTUGAL	1 milreis 1000 reis	4 6	A Set 4.444 milreis B To 1 £

Country.	Money of Account. Gold and Silver.	Average Current Value. s. d.	Comparison with the £ English. (*Any* Rate of Exchange may be set to the £ sterling in lieu of the following.)
PRUSSIA	11 thaler 30 silbergrochen 12 pfennige	3 0	A Set 6.20 thalers &S.G. B To 1 £
RUSSIA	1 ruble 100 copecs	3 2	A Set 6.25 ruble B To 1 £ .
SPAIN	1 dollar	4 2	A Set 4.16 dollars & reals B To 1 £
SWITZERLAND	1 franc 100 centimes	0 9½	A Set 25 francs B To 1 £
TURKEY	1 piaster 40 paras	0 2.1	A Set 115 piasters B To 1 £

WEIGHTS AND MEASURES.

AMERICA, U.S.

A Set 1 lb. } Weights similar.
B To 1 lb. }

Foot = 12 inches English

A Set 12 gallons, U.S.
B To 10 gallons, English

AUSTRIA

A Set 50 pfund
B To 62 lbs.

Foot = 12.445 in. Eng.

A Set 50 viertels
B To 156 gallons

BELGIUM Metre = 39.371 inches	The weights and measures of Belgium, France and Holland, are assimilated under the metrical system (see details under France), the nomenclature only differing, viz., Holland, the pond is the kilogramme and the kunn is the litre; Belgium, the livre is the kilogramme, and the litron is the litre.	

CHINA

Impl. foot = 12.162 inches

A Set 100 catties or pounds
B To 133 pounds weight

A Set 10 taus
B To 12 gallons

DENMARK

Foot = 12.357 inches

A Set 50 punds
B To 55 lbs.

A Set 50 viertels
B To 85 gallons

GERMANY

A Set 50 pfund
B To 53 lbs.
              Varying in States.

A Set 50 viertels
B To 80 gallons

ITALY

Florence foot = 11.94 inches
  „   Braccio = 22.978

A Set 50 rotollo
B To 52 lbs.

A Set 10 barile
B To 163 gallons

PORTUGAL

A Set 100 arratel or pound
B To 101 pounds

A Set 10 almude (Lisbon) 10 (Oporto)
B To 364 gallons   552

PRUSSIA	A	Set 50	pfund
	B	To 51	lbs.
Foot = 12.357 inches			
	A	Set 10	eimer
	B	To 151	gallons
RUSSIA	A	Set 50	pounds or fuut
	B	To 45	lbs.
Foot = 13.75 inches			
	A	Set 10	vedras
	B	To 27	gallons
SPAIN	A	Set 50	libra
	B	To 51	lbs.
Foot = 11.13 inches			
	A	Set 10	arobas or cantaros
	B	To 35	gallons
SWEDEN AND NORWAY	A	Set 50	skalpund
	B	To 48	lbs.
	A	Set 50	kanne
	B	To 29	gallons
SWITZERLAND	A	Set 50	pfund
Berne	B	To 57	lbs.
Foot = 11.54 inches			Varying in different Cantons.
	A	Set 10	eimer
	B	To 92	gallons
TURKEY	A	Set 50	rotolo
	B	To 63	lbs.
Pic = 26.8 inches			
	A	Set 100	almud
	B	To 115	gallons

Any foreign denomination may be so converted; having its value singly in English, multiply it say by 50, and having found its equivalent, set the *constants* on A　Set 50 of *any* kind foreign
　　　　　　　　　　　　　　　B　To = ? of given kind English.

COMPARISON AND CONVERSION OF FRENCH (METRICAL) AND ENGLISH STANDARDS IN DETAIL.

LENGTH.

Ex. : A   Set ? compare French standards on A
B   To ? with English standards   on B

A   Set 76   millimètres
B   To 3   inches

A   Set 76   centimètres
B   To 30   inches

A   Set 76   decimètres
B   To 25   feet

A   Set 55   metres (unit of length)
B   To 60   yards

A   Set 11   decamètres
B   To 120.3   yards

A   Set 11   hectomètres
B   To 1203   yards

A   Set 50   kilomètres
B   To 31   miles

A   Set 50   myriamètres
B   To 310.7   miles

Note.—The metre is the unit of length, and = 39.3708 inches, or 3.2809 feet.

SOLID.

A   Set 17   decistères
B   To 60   cube feet

A   Set 23   stères (unit = cube metre = 35.314
B   To 30   cube yards   cube feet.)

A   Set 23   decastères
B   To 300   cube yards

WEIGHT.

```
A   Set 1000 milligrammes
B   To  15.4 grains

A   Set  100 centigrammes
B   To  15.4 grains

A   Set   10 decigrammes
B   To  15.4 grains

A   Set  100 grammes (unit) 15.434 grains troy
B   To  1544 grains

A   Set   20 decagrammes
B   To    7. ounces avdp.

A   Set   50 decagrammes
B   To  16.1 ounces troy

A   Set   20 hectogrammes
B   To  70.5 ounces avdp.

A   Set   50· kilogrammes = 2 lb. 3 oz. 4½ dr. avdp.
B   To  110.3 lbs. avdp.

A   Set   10 myriagrammes
B   To  220.6 lbs.  avdp.

A   Set   50 quintals = 1 cwt. 3 qrs. 24½ lbs.
B   To   98.4 cwts.

A   Set   61 millier bar = 9 tons. 16 cwt. 3 qrs. 12½ lbs.
B   To   60 tons
```

SOLID.

```
A   Set  410 cube centimètres
B   To   25 cube inches

A   Set   85 cube decimètres
B   To    3 cube feet

A   Set   23 cube metres = 35.314 cubic feet
B   To   30 cube yards
```

CAPACITY.

A Set 1000 millitres
B To   61 cubic inches

A Set 100 centilitres
B To   61 cubic inches

A Set ⋅ 10 decilitres
B To   61 cubic inches

A Set   50 { litres, unit of liquid capacity.
               { a cubic decimètre $= 61.02379$ cubic ins.
B To   88 pints

A Set   5 decalitres
B To   11 gallons

A Set   2 hectolitres
B To   44 gallons

A Set   2 kilolitres
B To   440 gallons (1 kilolitre $= 35.3147$ cubic feet)

A Set   23 myrialitres
B To   301 cubic yards

SURFACE.

A Set   50 centiares
B To   60 square yards

A Set   10 ares (unit) {square decamètre
                          {$= 119.5991$ square yards
B To 1196 square yards

A Set   20 ares
B To   80 square rods

A Set   49 decares
B To   12 square acres

A Set   30 hectares
B To 74·2 acres

AVOIRDUPOIS.

A Set 53 grammes
B To 30 drachms

A Set 85 grammes
B To   3 ounces

A Set 50 kilogrammes
B To 110·3 pounds

TROY.

A Set 26 grammes
B To 400 grains

A Set 31 grammes
B To 20 dwts.

A Set 311 grammes
B To 10 ounces

A Set 30 kilogrammes
B To 80 lbs.

SYSTÈME USUEL.

A Set 100 pieds & Set 10 toises
B To 109 feet   To 66 feet

A Set 10 livres, 500 grammes each
B To 11 lbs. avdp.

A Set 20 livres
B To 27 lbs. troy

A Set 8 lieues poste
B To 5 miles

REDUCTION OF QUANTITIES AND PRICES (on A and B).

Ton.
A Set 5 Below £ cost per ton.
B To 5 Find Shillings per cwt.
A Set 28 Below £ cost per ton.
B To 3 Find Pence cost per lb.

	A	Set 28	Below	£ cost per cwt.
	B	To 5	Find	Shillings per lb.
	A	Set 7	Below	£ cost per cwt.
	B	To 15	Find	Pence per lb.
	A	Set 28	Below	Shillings cost per cwt.
	B	To 3	Find	Pence per lb.
Cwt.	A	Set 4	Below	£ cost per cwt.
	B	To 10	Find	Shillings per stone 14 lbs.
	A	Set 8	Below	Shillings cost per cwt.
	B	To 12	Find	Pence per stone 14 lbs.
	A	Set 7	Below	£ cost per cwt.
	B	To 10	Find	Shillings per stone 8 lbs.
	A	Set 7	Below	Shillings cost per cwt.
	B	To 6	Find	Pence per stone 8 lbs.
Lb.	A	Set 4	Below	£ cost per lb.
	B	To 5	Find	Shillings per oz.
	A	Set 20	Below	Shillings cost per lb.
	B	To 15	Find	Pence per oz.
Score.	A	Set 20	Below	Shillings per score.
	B	To 12	Find	Pence each
per 100	A	Set 25	Below	Shillings per 100
	B	To 3	Find	Pence each
pr.stone 14 lb.	A	Set 7	Below	Shillings per stone 14 lbs.
	B	To 6	Find	Pence per lb.
pr.stone 8 lb.	A	Set 4	Below	Shillings per stone 8 lbs.
	B	To 6	Find	Pence per lb.
Gross, 144.	A	Set 60	Below	Shillings per gross (144)
	B	To 5	Find	Pence each

E

Acre.	A	Set	8	Below	£ cost per acre.
	B	To	1	Find	Shillings per square rod.
	A	Set	4	Below	£ cost per acre.
	B	To	6	Find	Pence per square rod.
	A	Set	123	Below	£ cost per acre.
	B	To	6	Find	Pence per square yard.

Square rod.	A	Set	20	Below	Shillings cost per square rod.
	B	To	8	Find	Pence per square yard.

Load 40 bush.	A	Set	6	Below	£ cost per load 40 bushels.
	B	To	3	Find	Shillings per bushel.

Quarter	A	Set	40	Below	Shillings per quarter.
	B	To	5	Find	Shillings per bushel.

Sack 168 lbs.	A	Set	42	Below	Shillings per sack (168 lbs).
	B	To	36	Find	Pence per gallon (7 lbs.)

Sack 280 lbs.	A	Set	40	Below	Shillings per sack (280 lbs.)
	B	To	12	Find	Pence per gallon (7 lbs.)

Flour.	A	Set	45	Below	Shillings per sack.
	B	To	6	Find	Pence per 4 lb. loaf.

Flour.	A	Set	70	Below	lbs. of flour used
	B	To	90	Find	lbs. of bread made.

Barrel 36 gals.	A	Set	12	Below	Shillings per barrel (36 gals.)
	B	To	1	Find	Pence per quart.

£ per year.	A	Set	55	Below	£ expended per year.
	B	To	3	Find	Shillings per day.
	A	Set	35	Below	£ per year.
	B	To	23	Find	Pence per day.
	A	Set	13	Below	£ per year.
	B	To	5	Find	Shillings per week.

TROY AND AVOIRDUPOISE REDUCED (on A and B).

					Grains.	
A	Set	35	Below	Troy dwts.	of 24	each.
B	To	31	Find	Avdp. drachms	27.344	,,
A	Set	55	Below	Troy ounces	480	,,
B	To	60	Find	Avdp. ounces	437.5	,,
A	Set	204	Below	Troy ounces	480	,,
B	To	14	Find	Avdp. lbs.	7000	,,
A	Set	175	Below	Troy lbs.	5760	,,
B	To	144	Below	Avdp. lbs.	7000	,,
A	Set	6	Below	Troy lbs.	5760	,,
B	To	79	Find	Avdp. oz.	437.5	,,

JEWELLERS' WEIGHTS.

A	Set	4	Below	Troy grains
B	To	5	Find	Carat grains

GOLD, SILVER, AND COINAGE (on A and B.)

A	Set	39	Below Mint price in pence for standard gold, 934.5 pence per oz.
B	To	20	Find number of grains weight
A	Set	20	Below shillings per oz. troy
B	To	12	Find pence price per dwt.
A	Set	7	Below shillings per oz. troy
B	To	2	Find the carats fine, &c.
A	Set	93	Bank price in £ for standard gold
B	To	24	Number of oz. troy

WEIGHT IN COINS.

A	Set	67	Dwts. troy	Set 9 oz.	Set 3 lbs. troy
B	To	13	No. of sovs.	To 35 sovs.	To 140 sovs.
A	Set	40	Dwts. troy	Set 2 Oz. troy	
B	To	11	Shillings	To 11 Shillings	
A	Set	1	Oz. avdp.	Set 1 lb. avdp.	
B	To	5	Bronze halfpence	To 80 Bronze halfpence.	

E 2

SUPER. MEASURE, *whole price* in Shillings found (on A and B).

Dimensions, feet long, and feet broad.
   ,,   inches ,,   inches ,,

The following method of pricing supers. will be found very expeditious and useful.

Price per foot or yard.	No. 1. Gauge Points.	Price.	No. 2. Gauge Points.	Price.	No. 3. Gauge Points.
1	240·	1	12·	1	1728·
1½	160·	1½	8·	1½	1150·
2	120·	2	6·	2	864·
2½	96·	2½	4·8	2½	690·
3	80·	3	4·	3	576·
3½	68·4	3½	3·42	3½	492·
4	60·	4	3·	4	432·
4½	53·2	4½	2·66	4½	362·
5	48·	5	2·4	5	345·6
5½	43·6	5½	2·18	5½	315·
6	40·	6	2·	6	288·
6½	37·	6½	1·85	6½	266·
7	34·2	7	1·71	7	246·24
7½	32·	7½	1·6	7½	230·
8	30·	8	1·5	8	216·
8½	28·2	8½	1·41	8½	204·
9	26·6	9	1·33	9	192·
9½	25·2	9½	1·26	9½	182·
10	24·	10	1·2	10	172·8
10½	22·8	10½	1·14	10½	163·
11	21·8	11	1·09	11	156·
11½	20·8	11½	1·04	11½	150·
12	20·	12	1·	12	144·
15	16·	15	·8	15	115·2
18	13·2	18	·66	18	96·
21	11·42	21	·571	21	82·32
24	10·	24	·5	24	72·
30	8·	30	·4	30	57·6
36	6·66	36	·333	36	48·
42	5·72	42	·286	42	41·6

Column 1 note (No. 1): Below the other dimension, *same kind,* Find the whole price in £. — A To the G P. B Set one dimension.

Column 2 note (No. 2): Below the other side in yds. or feet, *same kind,* Find *whole price* of super. in shillings at the price per square yard or foot. — A To G P. B Set one side in yards or feet.

Column 3 note (No. 3): Below the other side in inches Find *whole price* of super. in shillings at the price per square foot. — A To G P. B Set one side in inches.

Ex.: What will 20 × 40 cost at 7½d. per square yard ?

	G	P	
A	32	40	
B	20		£25
			answer.

Ex.: What will 15 × 9 cost at 2½d. per foot ?

	G	P	
A	4·8	15	
B	9		28s. 1½d.
			answer.

Ex.: What will 50 × 40 inches come to at 7½d. per square foot ?

	G	P	
A	230	50	
B	40		8s. 8d.
			answer.

SIMPLE INTEREST FOR DAYS, WEEKS, MONTHS, AND YEARS,

At any rate per cent., and for any principal.

The most unwieldly Interest Tables published are limited, compared with the range of the Slide Rule in working such questions; as a check upon such calculations its assistance is invaluable, for although, owing to the minuteness of the divisions on so small an instrument, the lower fractional parts are sometimes difficult to read, the correctness of the work is at once tested without labour or a chance of error. Almost any rate may be obtained by doubling the *result* of some one of the following, as twice the answer of 3½ = 7 per cent., &c. Or by *halving* the gauge point for any rate given below, we find a new one for double that rate.

General Formula.

A	To the gauge point for rate	Below the principal
B	Set the given time	Find the interest (see examples).

Interest shown in	Shillings	Shillings	Shillings	£	
Rate per cent.	**Gauge Point for days. No. 1.**	**Gauge Point for weeks. No. 2.**	**Gauge Point for months. No. 3.**	**Gauge Point for years. No. 4.**	**Rate per cent.**
2	910·	180·	30·	50·	2
2 ½	730·	104·	24·	40·	½
3	602·	86·5	20·	33·2	3
3 ⅛	580·	83·	19·2	32·	
3 ¼	561·	80·	18·5	30·7	¼
3 ⅜	540·	77·2	17·8	29·6	⅜
3 ½	522·	74·2	17·1	28·	½
3 ⅝	503·	72·	16·5	27·6	⅝
3 ¾	486·	69·2	16·	26·6	¾
3 ⅞	470·	67·5	15·5	25·6	⅞
4	455·	65·	15·	25·	4
4 ⅛	442·	63·5	14·5	24·2	⅛
4 ¼	430·	61·1	14·1	23·4	¼
4 ⅜	417·	59·5	13·6	22·8	⅜
4 ½	405·	57·7	13·3	22·2	½
4 ⅝	395·	56·5	12·9	21·6	⅝
4 ¾	384·	55·5	12·6	21	¾
Interest shown in	Shillings	Shillings	Shillings	£	

Interest shown in	Shillings	Shillings	Shillings	£	Interest shown in
Rate per cent.	Gauge Point for days. No. 1.	Gauge Point for weeks. No. 2.	Gauge Point for months. No. 3.	Gauge Point for years. No. 4.	Rate per cent.
4 $\frac{7}{8}$	375·	53·2	12·2	20·46	$\frac{7}{8}$
5	365·	52·	12·	20·	5
5 $\frac{1}{8}$	356·	51·	12·7	19·6	$\frac{1}{8}$
5 $\frac{1}{4}$	347·	42·4	11·4	19·	$\frac{1}{4}$
5 $\frac{3}{8}$	340·	48·7	11·2	18·7	$\frac{3}{8}$
5 $\frac{1}{2}$	332·	47·3	10·9	18·2	$\frac{1}{2}$
5 $\frac{5}{8}$	324·	46·4	10 6	17·8	$\frac{5}{8}$
5 $\frac{3}{4}$	317·	45·1	10·4	17·4	$\frac{3}{4}$
5 $\frac{7}{8}$	310·	44·2	10·2	17·1	$\frac{7}{8}$
6	304·	43·2	10·	16 6	6
6 $\frac{1}{8}$	298·	42·3	9·8	16·3	$\frac{1}{8}$
6 $\frac{1}{4}$	292·	41·5	9·6	16·	$\frac{1}{4}$
6 $\frac{3}{8}$	286·	40·8	9·4	15·7	$\frac{3}{8}$
6 $\frac{1}{2}$	280·	40·	9 22	15·4	$\frac{1}{2}$
6 $\frac{5}{8}$	276·	39·2	9·1	15·1	$\frac{5}{8}$
6 $\frac{3}{4}$	271·	38·5	8·9	14 8	$\frac{3}{4}$
6 $\frac{7}{8}$	265·	37 7	8·7	14·5	$\frac{7}{8}$
7	260·	37·	8·55	14·24	7
7 $\frac{1}{8}$	256·	36 5	8 42	14·	$\frac{1}{8}$
7 $\frac{1}{4}$	252·	35·8	8·3	13·8	$\frac{1}{4}$
7 $\frac{3}{8}$	247·	35·2	8·2	13 5	$\frac{3}{8}$
7 $\frac{1}{2}$	243·	34·6	8	13·32	$\frac{1}{2}$
7 $\frac{5}{8}$	239·	34·0	7·9	13	$\frac{5}{8}$
7 $\frac{3}{4}$	235·	33 5	7·7	12·8	$\frac{3}{4}$
7 $\frac{7}{8}$	231·	33·	7·6	12·7	$\frac{7}{8}$
8	227·	32·5	7·5	12·44	8
8 $\frac{1}{8}$	224·	32·4	7 4	12·3	$\frac{1}{8}$
8 $\frac{1}{4}$	222·	31·5	7·26	12·1	$\frac{1}{4}$
8 $\frac{3}{8}$	219·	31·	7·2	11 9	$\frac{3}{8}$
8 $\frac{1}{2}$	215·	30·5	7·06	11·76	$\frac{1}{2}$
8 $\frac{5}{8}$	212·	30·	6 95	11 6	$\frac{5}{8}$
8 $\frac{3}{4}$	208·	29·6	6 85	11·4	$\frac{3}{4}$
8 $\frac{7}{8}$	205·	29·8	6 8	11 3	$\frac{7}{8}$
9	203·	28·75	6 66	11·1	9
9 $\frac{1}{8}$	200·	28 5	6 6	10 95	$\frac{1}{8}$
9 $\frac{1}{4}$	198·	28·	6·5	10 8	$\frac{1}{4}$
9 $\frac{3}{8}$	195·	25 7	6·4	10 65	$\frac{3}{8}$
9 $\frac{1}{2}$	192·	27 3	6·32	10 5	$\frac{1}{2}$
9 $\frac{5}{8}$	190·	27	6 25	10 4	$\frac{5}{8}$
9 $\frac{3}{4}$	187·	26·6	6·16	10 23	$\frac{3}{4}$
9 $\frac{7}{8}$	185·	26·3	6 1	10 1	$\frac{7}{8}$
10	182·	26·	6·	10·	10
Interest shown in	Shillings	Shillings	Shillings	£	

Ex. (from Column 1): What is the interest on £48 for 55 days
at $3\frac{7}{8}$ per cent.

A   G P 470 $\bigg\}$ for $3\frac{7}{8}$ £48
B   55   5.6*s.* = 5*s.* $7\frac{1}{4}d.$

The advantage of the Slide Rule is that the interest on *all* sums
is shown for the number of days once set to G P for any rate,
thus: for the above rate

A $\bigg\}$ Set   470   Below £30   £43   £60
B   55   Find 3.5*s.*   5*s.*   7*s.*

Ex. (from Column 2 for Weeks): What is the interest on £65 for
40 weeks at $3\frac{1}{4}$ per cent.

A   £65   '   Set 72 G P for $3\frac{1}{4}$ per cent.
B   36.11   To 40 weeks
        $1\frac{1}{2}$

Ex. (from Column 3 for Months): What is the interest on £55 for
9 months at $3\frac{1}{8}$ per cent.

A   Set   19.2 G P for $3\frac{1}{8}$ per cent.   £55
B   To   9 months   25.78
                                       $9\frac{1}{2}$

To those who adopt the Slide Rule for interest calculations it is
recommended to affix to this page decimal tables of weeks, months
and days, which can be formed on the Rule and copied easily, as

A   Decimals on A   1 Set
B   To days in a week on B   7 To

A TABLE SHOWING THE NUMBER OF DAYS, FROM ANY DAY IN THE
MONTH TO THE SAME DAY IN ANY OTHER MONTH.

	To	Jan.	Feb.	Mar.	April.	May.	June.	July.	Aug.	Sept.	Oct.	Nov.	Dec.
From	January	365	31	59	90	120	151	181	212	243	273	304	334
	February	334	365	28	59	89	120	150	181	212	242	273	303
	March....................	306	337	365	31	61	92	122	153	184	214	245	275
	April	275	306	334	365	30	61	91	122	153	183	214	244
	May	245	276	304	385	365	31	61	92	123	153	184	214
	June	214	245	273	304	335	365	30	61	91	122	153	183
	July	184	215	243	274	304	335	365	31	52	92	123	153
	August	153	184	212	243	273	304	334	365	31	61	92	122
	September..............	122	153	181	212	242	273	303	334	365	30	61	90
	October	92	123	151	182	212	243	273	305	335	365	31	61
	November	61	92	120	151	181	212	242	273	304	334	365	31
	December	31	62	90	121	151	122	212	243	274	304	335	365

COMPOUND INTEREST.

Rate per cent.

Years.	3.	4	4½	5	5½	6	Years.
1	1·0300	1·0400	1·0450	1·05	1·055	1·06	1
2	1·0609	1·0816	1·0921	1·1025	1·1131	1·1236	2
3	1·0927	1·1248	1·1411	1·1576	1·1742	1·191	3
4	1·1255	1·1698	1·1925	1·2115	1·2390	1·2625	4
5	1·1592	1·2166	1·2462	1·2763	1·307	1·3382	5
6	1·1940	1·2663	1·3023	1·3401	1·3788	1·4185	6
7	1·2298	1·3159	1·3609	1·4071	1·4547	1·5033	7
8	1·2668	1·3685	1·4221	1·4771	1·535	1·594	8
9	1·3047	1·4233	1·4861	1·5513	1·6191	1·6895	9
10	1·3439	1·4802	1·5530	1·6289	1·7082	1·791	10

(Constants)

By Slide Rule.

Ex. 1st. What will £550 amount to at compound interest in 4 years at 5 per cent.

A   To 1.211   Find 668.5 £ amount required
B   Set 1   Above 550

If for any term beyond the table, say 24 years, take some year by which it is divisible, and follow the series on the Rule, thus:

Ex. 2nd. £500 for 24 years at 5 per cent.

A   1.477   £738.5 (8 ys.)   1091.4 (16 ys.)   1612.5 (24 ys. required)
B   1   500   738.5   1091.4

Ex. 3rd. What sum will amount to £432 in 8 years at 3 per cent. compound interest.

A   To 1.267 (3 per cent. 8 ys.)   Below £432
B   Set 1   Find £341 sum required

The limits of such a work will not permit the application of the rule to the wide range of financial operations, which would occupy a volume; enough has been shown to prove the value of its assistance to the accountant. For the conversion of stock, exchanges, annuities, continuous and terminable, insurance, &c. &c., the saving of labour will be appreciated, especially in the absence of tables, logarithms,

&c., as a few constants which may be pencilled on the back of the rule itself suffice for particular questions.

———

PROPORTIONAL PARTS (on A and B).

The great number of *pro rata* questions that occupy the attention of commercial men must render any expeditious method of solution welcome; the peculiar applicability of the Slide Rule to such operations as dividends on shares, profits on separate advances, &c. &c., can easily be tested, and by its use proportions of one or many parts can be *instantly* found without any arithmetical process.

Ex.: Divide £80 in proportions of 2, 3, 5 = 10.

A	Below	2	3	5 parts		Set 10
B	Find	16	24	40	£	To £80

Ex.: Find the dividends at 8 per cent. on shares costing £35, 45, 62.10, and 70.

A	Below £35	£45	£62.10	£70	Set 100
B	Find £2.8	3.6	5	5.6	To 8 per cent.

Ex.: The assets being £120 and the debts £140, 80, 75, 50 = 345, find the respective dividends.

A	Below £50	75	80	140 debts	Set £345
B	Find £17.25	26	27.75	49 divds.	To 120

Further examples either of the simplicity or expedition of instrumental proportions are unnecessary.

———

PART VI.

SCIENTIFIC READINGS BY SLIDE RULE.

CONVERSION OF THERMOMETRIC SCALES (on A and B).

1. REAUMUR TO FAHRENHEIT.

A — Set 3   Read Reaumur on A   (*Note.* After complying with
B 1st To 9   Read Fahrenheit on B   (the conditions.

Above zero, Reaumur.   *Add* 14.2 to the given degrees.

Ex.: Convert 20° Reaumur to Fahrenheit.

A   Set   4   20 add 14.2 = 34.2
B   To   9   Answer 77° Fahrenheit

Under zero, Reaumur, and } Use the *difference* between the
   less than 14.2° given   }   given degrees, and 14.2

Ex.: Convert $\overline{7.1}$ Reaumur to Fahrenheit

A   Set   4   14.2 less $\overline{7.1}$ 7.1 the difference
B   To   9   Answer 16° Fahrenheit

Under zero, Reaumur, and } *Deduct* 14.2 from the given
   more than 14.2° given   }   degrees.

Ex.: Convert $\overline{16°}$ Reaumur to Fahrenheit.

A   Set   4 $\overline{16°}$ less 14.2 = 1.8
B   To   9   Answer $\overline{4°}$ Fahrenheit

2. Fahrenheit and Reaumur.

A	Set	4	Read Reaumur on A	*Note.* After complying with the conditions.
B	To	9	Read Fahrenheit on B	

Above zero, Fahrenheit { Find, and use the *difference* between the given degrees and 32

Ex.: Convert 50° Fahrenheit to Reaumur.

A	Set	4	Answer 8° Reaumur
B	To	9	50 less 32 = 18, the *difference*

Below zero, Fahrenheit. Add 32 to the given degrees

Ex.: Convert $\overline{10°}$ under 0 Fahrenheit to Reaumur.

A	Set	4	Answer $18\frac{2}{3}$° Reaumur
B	To	9	10 add 32 = 42, the *sum*

3. Centigrade and Fahrenheit.

Note.—The method is precisely as with Reaumur, the constants only differing.

A	1st.	Set	5	Read Centigrade on A	*Note.* After complying with the conditions.
B		To	9	Read Fahrenheit on B	

Above zero, Centigrade. Add $17\frac{7}{9}$ to given degrees

A	Set	5	Below given degrees? added to 17.7 (sum)
B	To	9	Find the degrees of Fahrenheit (?)

Under zero, Centigrade, and *less* than $17\frac{7}{9}$ } Use the *difference* between the given degrees and $17\frac{7}{9}$.

A	Set	5	Below the *difference* ($17\frac{7}{9}$, *less* the given degrees)
B	To	9	Find the degrees (*above* zero) Fahrenheit.

Under zero, Centigrade, and *more* than $17\frac{7}{9}$ } *Deduct* $17\frac{7}{9}$ from the given degrees.

A	Set	5	Below the given degrees? less $17\frac{7}{9}$?
B	To	9	Find the degrees (*below* zero) Fahrenheit?

4. FAHRENHEIT AND CENTIGRADE.

Note.—The method as with Fahrenheit and Reaumur, the constants only changed.

A   Set   5   Read Centigrade on A   ⎰ *Note.* After complying with
B   To   9   Read Fahrenheit on B   ⎱   the conditions.

Above zero, Fahrenheit ⎰ Find, and use the *difference* between the
                ⎱   given degrees and 32.

A   Set   5   Find the degrees of Centigrade ?
B   To   9   Above the *difference* between the given degrees and 32.

Below zero, Fahrenheit   *Add* 32 to the given degrees

A   Set   5   Find the degrees of Centigrade   ?
B   To   9   Above the given degrees *added* to 32   (sum).

5. CENTIGRADE AND REAUMUR.

Compared scales of Centigrade and Reaumur may be at once formed complete, without the conditions necessary to convert either to Fahrenheit.

A   Set   5   ⎰ Below degrees Centigrade on A ⎱ or *vice*
B   To   4   Then ⎱ Read degrees Reaumur   on B ⎰ *versâ*.

SPECIFIC GRAVITY (on A and B).

TO FIND THE SP. GR. OF ANY BODY.

A Set the weight lost in water   Below 1000· sp. gr. of water
B To the whole weight.   Find ?   sp. gr. of substance.

Ex.: A cubic foot of cast iron weighing 454 lbs., loses 62.5 lbs. when weighed in water. Required its sp. gr.

A   Set   62.5 lbs. loss in water   Below   1000
B   To   454   „   whole weight   Find   7271 sp. gr.

Having the contents in *cubic feet*, and sp. gr. of substance, to find the weight in lbs. avdp.

A   Set   7271 sp. gr. cast iron.   Find   454 lbs. avdp.
B   To   ·016 constant   Above (say)   1 cubic foot.

Having the contents in *cubic inches*, and sp. gr. of substance, to find the weight in lbs.

| A | Set | 7271 sp. gr. cast iron. | Find | 454 lbs. |
| B | To | 27·73 constant | Above (say) 1728 cubic inches. |

UNIFORM MOTION (on A and B.)

A	Set	1	Below miles per minute
B	To	88	Find feet per second.
A	Set	15	Below miles per hour
B	To	22	Find feet per second.
A	Set	5	Below miles per hour
B	To	440	Find feet per minute.
A	Set	23	Below miles = 5280 feet
B	To	20	Find knots = 6075·8 ,,

ACCELERATED MOTION.

1st.—To find the space in feet fallen through in any given number of seconds by any heavy body (on C and D).

Note.—The whole space fallen through, is as the square of the seconds occupied in falling, × by 16·1.

Ex.: A stone dropped from a cliff is observed to be four seconds falling. Required the height.

| C | Set | 400 | Find 257 feet | and | Find height in feet |
| D | To | 5 | Above 4 seconds. | | Above *any* seconds descent· |

2nd.—To find the velocity acquired during any given second of a body's descent (on A and B).

Note.—The spaces passed through for *each* second are as any given second of the descent doubled, *less* 1 and × 16·1.

Ex. : Through what space in feet will a body fall during the seventh second of its descent.

$$7 + 6 = 13$$

| A | Set | 16 | Find 209 feet during the seventh second |
| B | To | 1 | Above 13, double of 7, less 1. |

3rd.—To find the velocity acquired at the end of the time due to the height from which a body falls.

Note.—The velocity acquired at the end of the time is direct as 1, 2, &c., and the seconds × 32·2.

Ex.: What velocity will a falling body acquire at the end of eight seconds?

A   Set   32   Find 257 feet velocity at end of that time.
B   To   1   Above 8 seconds.

PROPORTIONS OF SPHERES.

SURFACE.

C   Set   50   Find surface in square super
D   To   4   Above diameter of sphere.

SOLIDITY.

C   Set   Diam.   Find cubic contents
D   To   4.37   Above diameter of sphere.

SIDE OF EQUAL CUBE.

A   Set   4   Find side of equal cube
B   To   5   Above diameter of sphere.

LENGTH OF EQUAL CYLINDER.

A   Set   4   Find height of equal cylinder
B   To   6   Above diameter of sphere.

WEIGHT OF WATER IN LBS. AVDP. IN ANY SPHERE.

C   Set   Diam.   Find lbs. water contained
D   To   7.26   Above diameter of sphere in inches.

FORCE OF WIND ON PERPENDICULAR SURFACE.

C   Set   12.3   Find force in lbs. on each sq. ft.
D   To   50   Above miles per hour of current.

TRAVELLING OF SOUND.

A	Set	14	Below seconds after report
B	To	3	Find miles distant.
A	Set	8	Below seconds after report
B	To	3000	Find yards distant.

VIBRATIONS OF PENDULUMS.

| C | Set | 39.14 | Find inches length |
| D | To vibrations per minute. | | Above 60 |

or

| S (Inv^d.) | 39.14 | } Set | { Table of inches length on S Inv^d. |
| D | 60 | | {   ,, vibrations per minute on D. |

WATER.

Supply of Water from Rainfall (on A and B).

In Symon's Register of Rainfall for 1865, the following formula is given :—

$$\left.\begin{array}{c}\text{Gallons daily} \\ \text{supply}\end{array}\right\} = 40,000 \times \left.\begin{array}{c}\text{Area in} \\ \text{sq. miles}\end{array}\right\} \times \begin{array}{c}\text{Inches available} \\ \text{Annual rainfall.}\end{array}$$

By the Slide Rule, the supply from any amount of rainfall over any area is instantly tabled.

General Rule.

| A | To the G P .2.5 | Below *any* inches annual rainfall |
| B | Set the area in sq. miles. | Find a number, to which *add* 5 ciphers, showing the gallons daily supply. |

Ex. : From an available area of 21 square miles, with an annual rainfall of 15 inches, find the supply in gallons daily.

A	Set 2·5	Below 15 in. rainfall
		add 5 ciphers
B	To 21 sq. miles	Find 12,600,000 gallons daily.

Or for Rainfall per Acre :

| A | Set 8 | Below any inches annual rainfall. |
| B | To 500 | Find gallons daily per acre. |

FORCE OF WATER.   (C and D).

When the velocity is given in feet per second.

C Set 1   Find the force in lbs. per sq. ft. (deduct ₁⁄₁₀″ part.)
D To 1   Above the velocity in feet per second.

Ex.: The velocity being 9 feet per second, required the force in .bs. on an area of 6 square feet.

C Set 6 square feet.   486, less ₁⁄₁₀ = 474 lbs.
D To 1   9 feet.

Velocity given in miles per hour.

C Set 2.1   Find the force in lbs. per square foot.
D To 1   Above the velocity in miles per hour.

Ex.: Required the *whole* pressure in lbs. on an area of 6 square feet, velocity being 4 miles per hour.

C Set 2.1   33·3 lbs. per foot × 6 = 199.8 lbs.
D To 1   4

PRESSURE OF WATER.   (on A and B).
Vertical Pressure.

Lbs. per square foot.

A Set 500   Find the pressure per square foot in lbs.
B To 8   Above any depth in feet.

Any area in square feet, cwts. pressure, to depth in feet.

A Set the depth in feet   Find cwts. pressure
B To   1·8   Above any area in square feet.

Any area in square feet, *tons* pressure, to depth in feet.

A Set depth in feet   Find tons pressure
B To G P 36.   Above *any* area in square feet.

Ex.: The depth being 20 feet and area 100 square feet, find the number of tons pressure.

A   20   56 tons
B G P 36   100 square feet area.

Lateral Pressure.

A   Below ½ depth.   To 36
B   Find tons lateral pressure   1st. { Set   ? area of side, ft. × ft.

———

OVERSHOT WATERWHEELS.

To find the horse power.

1st. Cube the *radius* of the wheel in feet.

2nd. Find the square root of cubed number.

3rd. Find the square feet area in cross section of the stream.

(All very simple operations by the Slide Rule).

General Rule.

A   Find the horse power.   (Set the sq. root (No. 2).
B   Above square feet area (No. 3).   1st. { To   6.5   G P

Ex.: An overshot wheel is 30 feet diam., section of stream 6 square feet, required the horse power.

1st. C   Set   15 radius.   Find 3375 { square root of which
      D   To   1   Above 15   on D is 58.

Then:

A   Find 53.6 horse power.   Set   58 sq. root (No. 2).
B   Above 6 square feet area.   To   6.5 G P

———

PUMPS.

The handle of a pump is 60 inches long, the bucket arm is 21.5 inches, and the hand stroke 13.5 inches, required the bucket stroke.

A   Find   7.6 in. bucket stroke (*a*)   To   13.5 the hand stroke
B   Above 21.5 the bucket arm.   Set 38.5 differ. of length.

With a six-inch barrel, working at 20 feet depth, required the weight of water and gallons lifted.

C   Set   20 feet.   Find 24.5 gall. = 245 lbs.
D   To   5.4 G P   Above 6 in. diameter.

Power required to lift (same Example).

A   Find   123 lbs. power req^d.   Set   245 lbs. lift
S (Inv^d). Above 38.5 long arm.   To   21.5 bucket arm.

With a handle 60 inches in length, required the place of the fulcrum for a stroke of 15 inches, the *power* making 25 inches stroke.

To 25 add 15 = 40, sum of strokes.

A   Find (21.5 ins. Ful.)   Find   38.5   Set 60 the length.
B   Above 15   Above 25   To 40 the sum of strokes.

PUMPING ENGINES (on C and D).

To find the required diameter of a steam cylinder (at 10 *lbs.* effective pressure per inch) for a pump of *any* diameter, and at *any* yards deep.

Note.—The constants given for each diameter of pump, when once set on the Rule, form tables of yards deep on C,
and of diameter of steam cylinder on D, thus :

C   Set { The given constants } Below *any* No. of yards deep on C
D   { for any dia. of pump } Find the req^d. dia. of steam cyl. on D.

Ex.: Required the diameter of a steam cylinder (at 10 lbs. effective pressure) to work a 12-in. pump, 40 yards deep.

C   Set   22 ⎫   Below   40 yards deep.
D   To   20 ⎭   Find   27 in. diam. of steam cylinder required.

Diameter of pump in inches ...			3	4	5	6	7	8	9
Below any No. of yards deep... on	C	Set	87	50	32	22	10	49	25
Find diam. of steam cylinder on	D	To	10	10	10	10	25	20	16

Diameter of pump in inches ...			10	11	12	13	14	15	16
Below any No. of yards deep... on	C	Set	20	25	22	19	25	20	49
Find diam. of steam cylinder on	D	To	16	20	20	20	25	24	40

Diameter of pump in inches ...			17	18	19	20	21	22	23
Below any No. of yards deep... on	C	Set	50	20	15	40	22	20	30
Find diam. of steam cylinder on	D	To	43	27	25	43	35	35	45

Diameter of pump in inches ...			24	25	26	·27	28	29	30
Below any No. of yards deep... on	C	Set	22	20	25	20	40	15	14
Find diam. of steam cylinder on	D	To	40	40	46	43	63	40	40

Note.—The above constants are all calculated to 10 *lbs. per inch* effective pressure ; but when it is required to find at greater or less pressure than that, *invert* the Slide to D, and taking the example given above.

S (Inv^d). Set 10 lbs.   Below *any other* given pressure
D   To 27 diam. at 10 lbs. } Find the required diameter.
                   pressure. }

as

S (Inv^d). Set 10   Below (say) 7 lbs. pressure (same pump)
D   To 27*   Find 32.4 ins. req^d. diam. at 7 lbs.

WEIGHT AND VOLUME OF WATER (on A and B, and C and D.)

A Set 19   Below cubic inches of water
B To   10   Find   ounces troy.

A Set 160   Below cubic inches of water
B To   7 ·   Find   lbs. troy.

A Set 26   Below cubic inches of water
B To   15   Find   oz. avdp.

A Set 416   Below cubic inches of water
B To   15   Find   lbs. avdp.

A Set 97   Below cubic feet of water
B To   54   Find   cwts.

A Set 36   Below cubic feet of water
B To   1   Find   tons.

A Set 4   Below cubic yards of water
B To   3   Find   tons.

A Set 22   Below cylindrical inches of water
B To   10   Find   oz. avdp.

A Set 350   Below cylindrical inches of water
B To   10   Find   lbs. avdp.

| A | Set | 20 | Below cylindrical feet of water |
| B | To | 13 | Find   cwts. |

| A | Set | 46 | Below cylindrical feet of water |
| B | To | 1 | Find   tons. |

| A | Set | 30 | Below spherical inches of water |
| B | To | 15 | Find   oz. avdp. |

| A | Set | 160 | Below spherical inches of water |
| B | To | 3 | Find   lbs. avdp. |

| A | Set | 31 | Below spherical feet of water |
| B | To | 9 | Find   cwts. |

| C | Set | 34 | Below lbs. weight of water, 1 foot deep |
| D | To | 10 | Find   diameter of cylinder in inches. |

| C | Set | 8 | Below cubic feet, at 1 foot deep |
| D | To | 38 | Find   diameter of cylinder in inches. |

| C | Set | 6.5 | Below gallons in 3 feet length |
| D | To | 8 | Find   diameter of pipe in inches. |

| A | Set | 100 | Below gallons |
| B | To | 16 | Find   cubic feet. |

<hr>

CASK GAUGING (on C and D).

Casks are gauged as cylinders, to which they are approximated, by finding a mean between the head and bung diameters. There are several methods, but the following simple rule answers all practical purposes, viz. : To find the mean diameter.

Multiply the difference between the head and bung diameters by ·63, and add the product to the head diameter for a mean diameter, thus :—

General Rule.

C   Set the length in inches   Find the gallons content
D   To   * G P 18.79   Above the mean diameter, ins.

Ex. : A cask is 40.7 ins. length.

33.6 bung diameter ⎫
25.8 head diameter ⎬ Mean diameter 31.3 ins.

Difference   8.8 ×.63 = 5.5, *add* to head dia. 31.3

C   Set 40.7   Find 111 gallons content
D   To   18.79 G P   Above 31·3 mean diamr.

BUILDERS' TONNAGE OF SHIPS (on C and D).

C   Set the length in feet.†   Find tons
D   To   13.75   Above breadth of beam.

Ex. : A vessel is 300 feet long, 30 feet beam, required the builders' tonnage.

C   300 feet.   1400 tons.
D   13.75   30 feet beam.

RATIO, SPEED, AND POWER OF STEAM-SHIPS
(on A and B).

RATIO.

A   Set   immersed section   Find the ratio
B   To   the power.   Above the *cube* ‡ of the speed.

SPEED.

A   Set   immersed section   Below the ratio
B   To   the power.   Find a No., ³√ of which = speed.‡

POWER.

A   Set   immersed section.   Find the power
B   To   the ratio.   Above the speed.

* The square root of cylindrical inches in a gallon.
† Between perpendiculars.
‡ The cube and cube root can both be found on C and D.   See rule 6, page 10.

AGRICULTURAL.

LAND MEASURE (on A and B).

A great amount of labour is saved in casting areas by the use of the Slide Rule, as it answers for *all dimensions*, whether taken in chains, yards, feet, perches, &c.

General Rule.

A	To Gauge Point (divisor)	Below the breadth
B	Set the length	Find the contents.

Ex.: Required the area when 8.50 chains by 4.70.

A	Below 4.70 breadth	1st {	To 10 G P
B	Find 4 acres.		Set 8.50 length.

Note.—In the following forms the Gauge Points on the *right hand* are for Triangles, the whole base and perpendicular being given; the Gauge Points or common divisors are doubled for the area of triangles.

MEASURE IN CHAINS AND LINKS, TO FIND ACRES.

			For Triangles.
A	Set G P 10	Below chains broad	(G P 20)
B	To chains long	Find acres content.	

MEASURE IN PERCHES, TO FIND ACRES.

A	Set G P 160	Below perches broad	(320)
B	To perches long	Find acres.	

MEASURE IN YARDS, TO FIND ACRES.

A	Set G P 4840	Below yards broad	(9680)
B	To yards long	Find acres.	

MEASURE IN YARDS, TO FIND PERCHES.

For Triangles.

| A | Set G P 30.3 | Below yards broad | (60.5). |
| B | To yards long | Find perches. | |

MEASURE IN LINKS, TO FIND PERCHES.

| A | Set G P 625 | Below links broad | (1250) |
| B | To links long | Find perches. | |

MEASURE IN FEET, TO FIND PERCHES.

| A | Set G P 272.2 | Below feet broad | (544.5` |
| B | To feet long | Find perches. | |

MEASURE IN MILES LONG, YARDS WIDE, TO FIND ACRES.

| A | Set 11 | Below yards wide |
| B | To 4 | Find acres per mile length. |

| A | Set 1 | Below chains wide |
| B | To 8 | Find acres per mile length. |

Note.—By *inverting* the slide to A, a table of yards wide and yards long, equal to 1 acre area, is found.

| A | Set | 4 | Table | { | Yards wide on A | } | = 1 acre. |
| S Inv^d. | To | 1210 | | { | Yards long on S Inv^d. | | |

REDUCING SCALES.

| A Set 1 | Below chains and links | or | A Set 50 | Below links |
| B To 66 | Read feet. | | B To 33 | Read feet. |

| A Set 24 | Below links | or | A Decimal parts of a link | Set 1 |
| B To 190 | Read inches. | | B Value in inches. | To 7.92 |

DECIMAL VALUES OF RODS IN AN ACRE, AND *vice versa.*

| A | Line of decimal parts, as | .25 dec. | 1st | { | To 1 |
| B | Value in rods | 40 rods, &c. | | { | Set 160 rods. |

MALT GAUGING.

Instead of the old method of gauge points, requiring constant practice, fixed numbers are given, which make the proportions of each question *evident*, and they can be used without hesitation or chance of error. The contents are calculated to 10 *inches deep;* the result being multiplied by *one-tenth* of the actual depth in inches, the whole contents are instantly found.

Note.—Find the square inches in area of the cistern; if square, by multiplying length and breadth on A and B, then—

General Rule for SQUARE CISTERNS (on A and B).

A Set   9 bushels   Find the bushels to *each* 10 ins. deep
B To 2000 sq. inches area   Above *any* number of sq. *ins.* area.

Ex. : A square cistern of malt is 32 inches deep, its sides being 84 and 60 inches, find bushels.

$$84 \times 60 = 5040 \text{ inches area.}$$

A To   9   Find   22.72 bushels, at 10 inches deep
B Set 2000   Above 5040
                       and to multiply
A Set 3.2 { One-tenth of   Find 72.7 bushels, whole contents
             actual depth
B To 1   Above 22.72

The detail of the simplest operation *seems* long, while three or four seconds suffice to complete the above work by the Slide Rule.

When the length and breadth are given in *feet* and the depth in inches—

A Set 13 bushels at 10 in. deep   Find bushels to *each* 10 in. deep
B To 20 square *feet* area   Above *any* square *feet* area.

Ex.: A cistern 32 inches deep, sides 7 × 5 feet = 35 sq. ft.

A Set 13 bushels   Find 22.7 bushels × 3.2 = 72.7 bushels conts.
B To 20   Above 35 feet area given.

ROUND CISTERNS (on C and D).

Diameter and depth in *inches.*

C Set 10 bushels	Find bushels to each 10 inches deep
D To 53 in. dia.	Above *any* diameter in *inches.*

Ex.: A cistern is 32 inches deep, and 60 inches diameter; required contents.

C Set 10	Find 12.75 × 3.2 48 bushels
D To 53	Above 60 inches diameter.

With diameter in *feet,* and depth in *inches.*

C Set 25 bushels	To find bushels at 10 inches deep
D To 7 feet	Above *any* diameter in inches

WEIGHT OF HAY IN STACKS.

To find the weight of hay in a stack, the dimensions in feet must first be carefully taken, viz. :—

Square-sided stacks.	Length.	Breadth.	*True* height.
Round do.	Diameter or breadth half-way up, for mean.		do.

*For the true height, measuring as a solid, add one-third of the height from the eaves to the top, to that of the eaves from the ground, then

Length × breadth × true height = cube contents of square stack.

Ex.: A stack is 30 feet long,
20 feet broad,
10 feet to eaves
9 feet eaves to top

10
$3 = \frac{1}{3}$
___
ft. 13, true height.

___

Long. Broad. True height.
30 × 20 × 13 = 7800 cubic feet or 289 cubic yards contents.

F

	Cubic yards.		Cubic yards.		Cubic feet.
New hay	Ton = 26.6	Cwt. = 1.33	or		36
Settled hay	,, = 20.0	,, = 1.			27
Old hay	,, = 17.7	,, = .885			24

Having found the contents in cube yards or cube feet, by far the simplest method is to use a constant for proportion on A and B for the *weight*, viz., so many cubic feet or cubic yards of each kind equalling 10 tons ; the rest is then merely a question of quantity and weight.

SQUARE STACKS.

New Hay.

A Set 10 tons   10 tons   Find the weight in tons
B To 7200 cubic ft. or 266.6 cubic yds.   Above the contents in cubic yards or cubic ft.

Settled Hay.

A Set *10 tons   10 tons   Find the weight in tons
B To 5400 cubic ft. or 200 cubic yards.   Above the contents in cubic yards or cubic ft.

Old Hay.

A Set 10 tons   10 tons   Find the weight in tons
B To 4800 cubic ft. or 177.7 cubic yds.   Above the contents in cubic yards or cubic ft.

Ex.: A square-sided stack of *settled* hay is 30 feet long, 20 feet broad, true height 13 feet ; required tons weight.

$$30 \times 20 \times 13 = 7800 \text{ cubic feet.}$$

A
B 1st { Set   10 tons   Find   14.4 tons.
          To   5400* cubic feet   Above 7800 cubic feet.

ROUND STACKS.

Instead of employing the old method of Gauge Points, which present difficulties to those unaccustomed to their use, it is far easier to consider each round stack as a cylinder of some fixed length. When a constant is obtained, answering for *all* diameters, 10 *feet length* or *height* is assumed here, and for any *different height* the weight can be instantly found ; **say 10 feet high gives 25 tons**; true height 23 feet.
                                                    2.3 multiply by
                                                    ──────────
                                 57.5 tons for 23 feet high

New Hay.

Constants.

C Set 15 tons   7 tons   Find tons weight *each* 10 *feet high*

            or

D To 37 ft. dia.   80 ft. circ.   Above *any* diam. or circ. in feet.

Settled Hay.

Constants.

C Set 18* tons   15 tons   Find tons weight *each* 10 *feet high*

            or

D To 35 ft. dia.   100 ft. circ.   Above *any* diam. or circ. in feet.

Old Hay.

Constants.

C Set 20 tons   8 tons   Find tons weight *each* 10 *feet high*

            or

D To 35 ft. dia.   70 ft. circ.   Above *any* diam. or circ. in feet.

Ex.: A stack of settled hay is 40 feet mean diameter, 27 feet true height; required weight in tons.

C Set 18   Find 23.3 tons *each* 10 *feet. high* × 2.7 = 63 tons whole

                                         weight

D To 35   Above 40 feet diameter.

* Constants for Settled Hay.

WEIGHT OF LIVE CATTLE (on C and D).

————

Considering the carcase of an animal as a cylinder of flesh of an average gravity, a close approximation may be made to its saleable weight when killed. Rules called cattle gauges are used, but the following is perhaps the simplest method, being without chance of error. The condition of the animal must, of course, be taken into account; the following is for "good or fair;" experience alone can guide the allowance to be made.

IN STONES OF 8 LBS.

G P
C Set 15 stones of 8 lbs. to *each*   Find the weight in stones of 8 lbs.
       10 *inches* of length   for each 10 inches length

D To 80 inches* girth.   Above any girth in inches

Ex. : A bullock is 90 inches in girth, and 55 inches in length; required its weight in stones of 8 lbs.

C To 15   Find 19 stones for each 10 inches length × 5.5 =
                                          104 stones

D Set 80   Above 90 inches girth

IN STONES OF 14 LBS.

G P
C Set 4 stones of 14 lbs. to *each* 10   Find weight in stones 14 lbs.
       inches of length   for each 10 inches length
D To 55 inches girth   Above any girth in inches.

Ex : A bullock is 90 inches in girth, and 55 inches in length; required its weight in stones of 14 lbs.

C Set 4 stones of 14 lbs.   Find 10.8 stones 14 lbs. for each 10 ins.
                     length × 5.5 = 59.4 stones of 14 lbs.

D To 55 inches girth   Above 90 inches girth.

* This admits of easier multiplying by any length taken in inches than when the length is given in feet.

CONSTRUCTION.

THE Sliding Rule having no "*works in it,*" and there being nothing about its mechanism to account for its unerring action, many are puzzled by its performance. A brief explanation of the principle of construction is given; brief, because those acquainted with logarithms need none, having, I dare say, divined its secret, as a musician would read a scale, at sight; while others will derive little information, until they have extended their inquiries beyond the limits afforded here.

I can only describe how logarithms are employed on it. Each space occupied by 1 to 10 (A B C are double, D single lines) is supposed to contain 1000 or 10,000 parts, irrespective of its length; and so many of these parts are allotted to each number as is represented by the logarithm of that number, thus :—

				Occupies a space equal to				
No. 1 whose log. is		0				1	—	0
,, 2	,,	301	of the 1000 parts;	between	1 and 2			
,, 3	,,	477	,,	,,		1	,,	3
,, 4	,,	602	,,	,,		1	,,	4
,, 5	,,	698	,,	,,		1	,,	5
,, 6	,,	778	,,	,,		1	,,	6
,, 7	,,	845	,,	,,		1	,,	7
,, 8	,,	903	,,	,,		1	,,	8
,, 9	,,	954	,,	,,		1	,,	9
,, 10	,,	1000	,,	,,		1	,,	10

All divisions and subdivisions are logarithmic parts also, and expressed by log. numbers; in fact, we use the space (occupied by so many parts) engraved on the Rule, instead of printed numbers in a book, which mean the same quantity. To ensure accuracy, a dividing engine is used to lay down the scale.

Without entering into the history of logarithms, we have something, even in the few given above, by which we can test their uses on the Rule; for instance, take

Multiply these numbers { No. 2 whose log. is 301 } add these logs.
and 4 ,, 602

8 ,, 903 == log. of 8

Now by the instrument (multiplication case 1), it is evident that by projecting the 1 on the Slide B to 2 on the line A, thus

A	1		2	4	8
B	Slide	☞	1	2	4

we have in reality *added* the 301 parts between 1 and 2, to the place *first occupied* by the 4, which already held 602 spaces, making 903, and reaching 8 on A, therefore multiplying 4 (and *all other numbers* on the slide) by 2, the motion of the Slide, adding to or subtracting from its logarithmic *place*, in relation to the fixed numbers on A. This is the key to the working.

C }
D } form lines of squares and roots,

because the *space* allotted between each number on D, and *not* the *number* of parts, is exactly double that on C.

Were C made a triple line to D, cubes and cube roots, instead of squares and square roots, would be found by inspection ; and other powers and roots could be so represented.

There are various kinds of Sliding Rules adapted to different uses ; the principle is the same in all, the arrangements only differing to answer specific purposes ; but the simple form here recommended will be found most generally useful, and its use most easily acquired.

I am greatly indebted to many writers on the instrument, whose able works scarcely reach the public, both for information and valuable suggestions. In naming one (Viscount Gage), the tribute is as much due to his earnest advocacy and recommendation of instrumental calculation, as to my grateful remembrance of his kindness.

CONCLUSION.

I CANNOT close this work without remarking, that there is hardly a limit to the use of the Slide Rule in ordinary questions of business. So far from the subject being exhausted, even by the wide range of examples given, it should be looked upon only as an introduction to instrumental calculation. Any person mastering its use will find endless opportunities for its employment, when in addition to its universal applicability in proportional statements, questions such as:

$$x = \frac{a^2 \cdot b}{c} \text{ or } x \; \frac{a_3 \cdot b}{c \sqrt{}}$$

are *instantly* solved, and are capable of being applied to a variety of operations, to the saving of time and unnecessary labour. In working with the Slide Rule, after a little practice, every step in the process appears as consequent as the figures in a sum, and as easily read by an intelligent eye; and the reason how, and *why* numbers are multiplied, and divided, squared and cubed by it, is as clear as in arithmetical solutions. It is unjust to call its operations merely mechanical, as far as the operator is concerned; for the ingenuity necessary to a correct method of stating each question is in itself a course of teaching, demanding mental exertion, but exempting it from the drudgery of the process.

The great value of the Slide Rule appears in its use in ordinary work; its assistance in gauging, timber measuring, mensuration of superficies and solids, gives only examples of its capabilities, which may be applied with equal advantage to the instant formation of tables of proportions, the comparison or conversion of things differing in quantity or value; in fact, it has been pronounced by those best capable of employing it, and estimating its powers, as "a marvel of ingenuity and utility." (*Guthridge.*)

One advantage to the public in the present work cannot fairly be overlooked: the publishers have decided on issuing with each copy a neat, portable, and perfectly reliable though inexpensive instrument, answering all the purposes for calculating, of the article either costly or unwieldy, generally made. Attention to the examples given will illustrate its use; and as engineers, contractors, builders, architects, accountants, &c., have each their special subjects on which to employ it, so the practical mechanic will find in the work a store of valuable information, instructive and amusing. My chief aim has been to make the subject clear and its working easy. The application must rest with those who can appreciate a really useful instrument. It would be difficult to find one that so well repays the trouble *and pleasure* of investigation.

VIRTUE AND CO., PRINTERS, CITY ROAD, LONDON.

ELLIOTT, BROTHERS,

449, STRAND, LONDON,

MANUFACTURERS OF

𝔐athematical, 𝔒ptical, and 𝔓hilosophical 𝔍nstruments,

TO THE ADMIRALTY, ORDNANCE, EAST INDIA COMPANY,

BOARD OF TRADE, MILITARY COLLEGES, &c. &c.

From 30, Strand, and 5, Charing Cross.

Theodolites, Levels, Transit Instruments, &c.

	£	s.	d.
Four-inch plain Theodolite	16	16	0
Five-inch ditto ditto	18	18	0
Six-inch ditto ditto	21	0	0
Four-inch best Theodolite, divided on Silver ..	21	0	0
Five-inch ditto, ditto	25	4	0
Six-inch ditto, ditto	28	7	0
Seven-inch ditto, ditto	33	12	0
Six-inch ditto, ditto with 2 Telescopes ..	34	13	0
Seven-inch ditto, ditto ditto ..	39	18	0
Four-inch best Transit Theodolite	25	4	0
Ditto ditto, without Vertical Circle ..	21	0	0
Five-inch ditto	28	7	0
Ditto ditto, without Vertical Circle ..	25	4	0
Six-inch ditto	33	12	0
Ditto ditto, without Vertical Circle ..	28	7	0

Larger Theodolites and Levels made to Order.

Hoare's Box Slide Rule to accompany this work, 3s.

	£	s.	d.
Seven-inch best Transit Theodolite	38	17	0
Ditto ditto, without Vertical Circle ..	33	12	0
Four-inch Everest Theodolite	21	0	0
Five-inch ditto	25	4	0
Six-inch ' ditto	28	7	0
Seven-inch ditto	34	13	0
Eight-inch ditto	39	18	0
Drainage Level, with Stand..	4	4	0
Small Dumpy Level	6	10	0
ELLIOTT's Improved Dumpy Level, 10-inch ..	14	14	0
Ditto, ditto, with Compass	16	16	0
Ditto, ditto, 14-inch	15	15	0
Ditto, ditto, with Compass	17	17	0
Ditto, ditto, 18-inch	17	17	0
Ditto, with Compass	21	0	0
Ten-inch Y Level	9	9	0
Ditto ditto, with Compass	10	10	0
Fourteen ditto	12	12	0
Ditto ditto, with Compass	14	14	0
Circumferenters, four-inch	6	6	0
Ditto, five-inch	8	8	0
Ditto, six-inch	9	9	0
Ditto, six-inch, with Vernier	11	11	0
Lean's Dial, with Telescope and Arc	17	17	0
Prismatic Compass, in sling case	2	12	6
Ditto, with Azimuth Glass, ditto	3	3	0
Ditto, large size ditto	4	4	0
Ditto, ditto, with Silver Compass, ditto	5	5	0
Stand for ditto	1	11	6
Four-inch Metal Sextant, divided on Silver ..	9	9	0
Five-inch ditto, ditto, ...	10	10	0
Six-inch ditto, ditto, ...	11	11	0
Eight-inch Metal Sextant, divided on Silver ..	13	13	0
Ditto ditto, ditto, with double frame ..	16	16	0
Six-inch ditto, as supplied to Naval Cadets ..	8	8	0

	£	s.	d.
Quadrant	2	12	6
Ditto, with Telescopes .. .. £3 13s. 6d. to	6	6	0
Artificial Horizon .. .. .. £2 12s. 6d. to	4	14	6
Pocket Sextant, in sling case	4	4	0
Ditto, small size, in case	3	10	0
Ditto, with Telescope, ditto	4	14	6
Ditto, ditto and Supplementary Arc, ditto ..	5	15	6
Ditto, ditto ditto, large size ditto ..	6	16	6
Optical Square	1	1	0
Pentagraph, 24-inch	5	15	6
Ditto, 30-inch	6	16	6
Ditto, 36-inch	7	17	6
Ditto, 42-inch	8	18	6
Ditto, 48-inch	10	10	0
Eidograph	10	10	0
Station Pointers .. .. .. £6 6s. to	18	18	0
Transit Instrument	16	16	0
Ditto, with Iron Stand	21	0	0
Ditto, 24-inch, with Circular Brass Portable Stand	25	4	0
Ditto, 30-inch, with Iron Stand	45	0	0
Large Transit Instruments, for permanent Stations	73	10	0
Twelve-inch Azimuth, and Altitude Instruments..	95	0	0
Fifteen-inch ditto, with Microscopes	120	0	0
Cross Staff .. .. 10s. 6. to		16	0
Ditto, with Compass and Screw Staff £2 2s. to	2	12	6
Current Meter .. .. .. £3 3s. to	5	5	6
Six-inch Circular Protractor, with Tangent Screw adjustment	4	14	6
Six-inch ditto, divided on Silver, with Tangent Screw adjustment	5	15	6
Seven-inch ditto, ditto, ditto ..	6	16	6
Eight-inch ditto, ditto, ditto ..	7	17	6
Metal and Brass Circular and Semi-circular Protractors .. .. 10s. 6d. to	5	5	0

4

		£	s.	d.
Portable Sliding Station Staff, 14 feet		2	12	6
Ditto, ditto, Painted..		3	0	0
Ditto, 17-feet		3	13	6
Gravatt's Level Staff		3	3	0
Papers for Level Staves .. .. per foot			0	2
Ditto, in sets for 14-feet Portable Staff			3	0
Chain, 66-feet .. .. .. 10s. to			14	6
Ditto, extra strong wire, with three oval rings ..			18	0
Ditto, 25-feet			8	6
Ditto, 50-feet			14	0
Ditto, 100-feet		1	2	0
Patent Galvanized Chain, 66-feet			18	0
Standard Chains .. .. .. £3 3s. to		5	5	0
Arrows, set			2	0
Tapes, 33-feet			7	0
Ditto, ditto divided decimally			7	6
Ditto, 50-feet			8	6
Ditto, ditto divided decimally			9	6
Ditto, 66-feet			10	0
Ditto, ditto divided decimally			11	0
Ditto, 100-feet			15	0
Pocket Tapes .. .. 3s. to			10	6
Steel Standard Tape Measures .. £1 5s. to		3	3	0
Helicograph, or Spiral Instrument		3	3	0
Farey's Elliptograph		7	17	6
Opisometer, or Map Meter, for measuring curved lines			3	0
Card Protractors			2	6
Sang's Platometer, for Calculating areas on plans		10	10	0
Amsler's Planimeter ditto ditto £2 12s. 6d. to		3	3	0

(*Reduced List of Prices*, 1868.)

Drawing Instruments, &c.

	Brass.	Ger. Silver.
	£ s. d.	£ s. d.
Proportional Callipers, 12-inch	2 0 0	2 12 6
Ditto ditto 9-inch.. ..	1 14 0	2 2 0
Proportional Compass, 12-inch.. ..	2 10 0	3 0 0
Ditto ditto 9-inch.. ..	2 0 0	2 5 0
Ditto ditto 6-inch.. ...	1 5 0	1 10 0
Ditto ditto with adjustment	1 15 0	2 0 0
Whole and half ditto	18 0	1 1 0
Triangle ditto	18 0	1 1 0
Tube Compass	1 15 0	2 0 0
Ditto, Needle Points	1 18 0	2 4 0
Tube Beam Compass	2 0 0	2 5 0
Pocket Turn-in Compass	1 6 0	1 10 0
Ditto ditto, with handles	1 10 0	1 15 0
Ditto ditto, ditto, and Bars	2 2 0	2 5 0
Pillar Compass, with divided Sheath ..	1 10 0	1 15 0
Napier Compass	1 10 0	1 15 0
Pocket Dividers, with Sheath	7 0	8 6
Best double-jointed Compass	15 0	18 0
Ink Pencil, and Wheel Legs for ditto each	5 0	6 0
Lengthening Bar	5 0	6 0
Hair Divider	8 6	10 6
Plain Divider	5 0	6 6
Tate's Bow	1 1 0	1 5 0
Double-jointed Bowpen and Bowpencil, each	10 0	11 6
Ditto ditto, Needle Pointed .. each	10 6	12 6
Hair Bow Divider	8 0	10 0
Large Needle Bows .. .. each	13 0	15 0

	Brass.			Ger. Silver.		
	£	s.	d.	£	s.	d.
Spring Bowpen and Pencil .. each		8	0		8	0
Spring Divider		7	0		7	0
Ivory Handle Drawing Pen .. ..·		4	0		4	0
Ditto ditto, jointed		5	0		6	0
Ebony ditto		3	0		3	0
Double Pen		10	0·		10	0
Six Wheels and Box		6	0		6	0
Six Pens and one Handle, in Case ..	1	5	0	1	5	0
Tracer		2	6		2	6
Crow-Quill Holder		2	6		2	6
Needle Holder		2	6		3	0
Knife		1	6		1	6
Sector joined Compass, 4½, 5, or 6-inch		9	0		12	0
Divider		5	0		6	6
Ink and Pencil Legs		9	0		12	0
Bar..		4	6	·6	0
Bowpen and Pencil		10	0		12	0
College Compass		7	0		9	0
Ink and Pencil Legs		7	0		9	0
Bar..		3	6		4	6
Divider		3	6		4	6
Bowpen and Bowpencil		7	0		9	0
Steel Joint Compass, with Pen and Pencil Legs		7	0			
Divider		2	6			
Bowpen and Pencil		5	0			
Drawing Pen		2	0			
Case of Instruments as supplied to the Royal Military College, Woolwich..	2	10	0	3	0	0
Ditto as supplied to Sandhurst	2	2	0			
Ditto as supplied to Harrow School ..	3	0	0			
Ditto as supplied to King's College ..	2	12	6			

	Brass. £ s. d.			Ger. Silver. £ s. d.		
Case of Instruments as supplied to Eaton College	2	12	6			
Ditto for Schools	1	1	0			
Ditto as supplied to Board of Trade	0	6	6			
Ditto ditto	0	15	0			
Prize ditto ditto	1	5	0			
Architect's Case	3	6	0	4	0	0
Engineer's ditto .. £3 10s. to	6	16	6	10	10	0
Magazine Case of Instruments made of Brass, German Silver, or Silver, from	10	10	0	50	0	0
Marquois Scales, Boxwood					10	0
Ditto, Ivory				2	12	6
Ditto, Metal				3	13	6
Ivory Plotting Scales, 12-inch					10	0
Ditto ditto 6-inch					5	0
Ditto Offset Scales, 2-inch					2	0
Ditto Architect's Scales, 12-inch					12	0
Ditto ditto 6-inch					6	6
Ivory Protractor, 12-inch				1	10	0
Ditto ditto 6-inch					6	0
Ditto ditto Rolling					18	0
Ditto, Sector 6s. 0d. to				1	1	0
Ditto Parallel Rule, Brass Bars					4	6
Ditto ditto, German Silver Bars					5	0
Routledge's Slide Rule, with Book					9	0
Carret's Improved Ditto					10	0
Hawthorn's Improved Ditto					12	0
Bayley's Slide Rule					7	0
Ditto, Double Slide					14	0
Ditto Treatise on Slide Rule					6	0
Box Plotting Scales, 12-inch					3	0
Ditto Offsets					1	0
Ditto Architect's Scale					5	0
Ditto, Pocket Rules 2s. 6d. to					14	0

	£ s. d.	£ s. d.
Ebony Parallel Rules	per inch	0 4
Rolling Parallel Rules	do.	1 0
Ditto ditto, with Ivory Edges ..	do.	1 6
Steel Straight Edges..	per foot	5 0
Lance-wood T Squares .. .. .. do.		3 0
Ebony T Squares .. .. do.		3 6
Computing Scales		16 0
Drawing Boards, Tachet's Patent .. .. 13s. to		1 2 0
Triangles in Metal, Wood, and Ivory.		
French Curves		
Shipwright's Curves		1 15 0
Centrolinead, small size		1 8 0
Ditto, large size		1 11 0
Sets of Radii Curves	£1. 1s. 0d. to	5 5 0
Trammels	£2. 12s. 6d. to	5 5 0
Beam Compasses	£1 0s. 0d. to	5 5 0
Lalanne's Glass Slide Rule and Book		5 0

All the above Scales are Divided by a very Superior newly invented Dividing Engine.

Scales, Rules, and Tapes, made to all Foreign Measures, also English and Foreign Standard Measures.

Telescopes, Microscopes, &c.

		£ s. d.
Twelve-inch Achromatic Telescope..	..15s. to	1 11 6
Eighteen-inch ditto	..21s. to	2 2 0
Twenty-four-inch ditto	..42s. to	3 3 0
Thirty-six-inch ditto	..84s. to	5 5 0
Ditto in Case, with Brass Stand		8 8 0
Twelve-inch Portable Achromatic Telescope ..		1 11 6
Eighteen-inch Portable Achromatic Telescope ..		2 2 0
Twenty-four-inch ditto ditto		2 12 6
Twelve-inch Portable Achromatic Telescope, German Silver		2 2 0
Eighteen-inch ..ditto. ditto ditto..		2 12 6

	£	s.	d.
Twenty-four-inch Portable Achromatic Telescope, German Silver	3	3	0
Thirty-inch ditto ditto ditto ..	4	4	0
Military Telescope	3	18	0
Ditto, Regulation Pattern	3	8	0
Naval Telescope with Caps and String from 24s. to	6	6	0
Thirty-six-inch Achromatic Telescope, on Stand with Astronomical Eye Pieces	12	12	0
Forty-five-inch ditto ditto ditto, 2¾-in. aperture	25	4	0
Ditto ditto ditto 3¼-in. aperture	38	0	0
Ditto ditto ditto 4-in. aperture	75	0	0
Admiralty Contract Telescope, as supplied to the Navy	3	0	0
An Excellent Student's Telescope, on Brass Stand, in Case, 42-inches long, 2¾-in. aperture, with 3 Eye Pieces	15	15	0
Powerful Double Field Glass, much used for Yachting or as a Night Telescope £3 13s. 6d. to	4	14	6
Gardener's Microscopes	0	4	0
Compound ditto	1	1	0
Ditto ditto, with Achromatic Lenses £5 5s. to	30	0	0
Medical ditto .. .. £4 4s. to	10	10	0
Solar ditto .. .. £6 6s. to	20	0	0
Stereoscopes, with Diagrams .. £1 1s. to	5	5	0
Ditto, Revolving, with 50 selected Glass Views ..	22	0	0

Telescopes made of all sizes. Opera Glasses mounted in Buffalo, Horn, Ivory, Tortoise-shell, Pearl, &c., in every variety and fashion.

Magic Lanterns and Dissolving Views of great variety.

Spectacles mounted in Gold, Silver, Steel, and Shell, with Glasses, Pebbles, or Chamblant Glasses, which are highly approved of for Spectacles, Reading Glasses, &c., as they overcome all spherical abberation.

OPTICIANS TO THE OPHTHALMIC HOSPITAL.

Barometers, Philosophical Apparatus, &c.

	£ s. d.		£ s. d.
Pediment Barometers	1 11 6	to	5 5 0
Wheel ditto	1 11 6	to	8 18 6
Standard ditto	5 15 6	to	8 18 6
Marine ditto	2 12 6	to	7 17 6
Aneroid ditto	2 12 6	to	3 0 0
Mountain ditto	3 13 6	to	6 16 6
Ditto ditto, with Legs ..	10 10 0	to	13 13 0

Wollaston's Boiling-point Thermometer, for ascertaining heights 5 5 0

Thermometers for Baths, Hot-houses, &c., of every description.

Hydrometers and Saccharometers from 7s. 6d. to	4 4 0
Urinometers .. .. .. from 7s. 6d. to	12 0
Galvanic Machines	3 10 0
Stringfellow's Patent Pocket Galvanic Battery ..	1 1 0
Patent Pedometer	4 4 0
Perambulator	5 10 0

Air Pumps, Electrifying Machines, and all other Philosophical Apparatus. (*See Illustrated Catalogue.*)

Complete Sets of Photographic Apparatus, with Camera, for Portraits and Landscapes, Chemicals, &c., in case .. .. .. quarter-plate 8 18 6
Ditto ditto .. .. .. half-plate .. 15 15 0
Ditto ditto .. .. .. whole-plate 25 4 0

Sole Manufacturers and Agents for Richards' Patent Steam Engine Indicator 8 10 0

Drawing Paper plain and mounted, Tracing Paper, Colours, Globes, Drawing Boards, Curves, Black Mirrors, Camera Lucidas, Stands for ditto, Camera Obscuras, Walking Sticks with Rods.

Universal Horizontal, and Ring Sundials; Models of Diamonds; ELLIOTT's Drawing Pencils, Water Colours, &c.

Treatise on the Slide Rule, by the Rev. W. ELLIOTT, 1s. 6d.

Treatise on Mathematical Instruments, by MR. HEATHER, 1s.

The Practice of Engineering Field Work, by MR. HASKOLI, 20s.

Electro-Magnetism, 1s. 6d.

Experimental Chemistry, 1s. 6d.

Treatise on the Slide Rule, by W. H. BAYLEY, ESQ., 6s.

Treatise on the Double ditto, ditto, 3s.

Treatise on the Slide Rule, by HOARE, with Cardboard Rule, 3s.

The same with Box-wood Rule, 6s.

Treatise on the Steam Engine Indicator, by MR. PORTER, 4s.

DESCRIPTIVE ILLUSTRATED CATALOGUES,

Published by ELLIOTT, BROTHERS, successors to Watkins & Hill, late 5, Charing Cross, and 30, Strand, London. Price 6d. each.

OPTICAL INSTRUMENTS.

ELECTRICAL INSTRUMENTS.

MAGNETIC AND ELECTRO-MAGNETIC INSTRUMENTS.

VOLTAIC AND THERMO-ELECTRIC INSTRUMENTS.

HYDROSTATIC, HYDRAULIC, PNEUMATIC, AND ACOUSTIC INSTRUMENTS.

GENERAL ILLUSTRATED CATALOGUE, One Shilling.

ELLIOTT, BROTHERS,

SOLE MANUFACTURERS AND AGENTS FOR

RICHARDS' PATENT STEAM ENGINE INDICATOR.

TREATISE ON THE INDICATOR,
BY MR. PORTER, 4s.

These Steam Indicators received the Prize Medal at the Exhibition, 1862, and are allowed to be the best in use by all the most eminent Engineers.

NOTICE.

ELLIOTT, BROTHERS have much pleasure in stating that they have been enabled to make a great reduction in the prices of Drawing Instruments in this Catalogue, and at the same time add with great confidence that they have improved the construction and manufacture of their Instruments generally.

G. WITT, Printer, 7, Earl's Court, Leicester Square, W.C.

CATALOGUE

OF

RUDIMENTARY, SCIENTIFIC, EDUCATIONAL, AND CLASSICAL WORKS,

FOR COLLEGES, HIGH AND ORDINARY SCHOOLS, AND SELF-INSTRUCTION;

ALSO FOR

MECHANICS' INSTITUTIONS, FREE LIBRARIES, &c. &c.,

PUBLISHED BY

VIRTUE & CO., 26, IVY LANE,

PATERNOSTER ROW, LONDON.

。 THE ENTIRE SERIES IS FREELY ILLUSTRATED ON WOOD AND STONE WHERE REQUISITE.

The Public are respectfully informed that the whole of the late MR. WEALE'S *Publications, contained in the following Catalogue, have been purchased by* VIRTUE & Co., *and that all future Orders will be supplied by them at* 26, IVY LANE.

。 Additional Volumes, by Popular Authors, are in Preparation.

AGRICULTURE.

66. CLAY LANDS AND LOAMY SOILS, by J. Donaldson. 1s.
140. MODERN FARMING: Soils, Manures, and Crops, by R. Scott Burn. 2s.
141. MODERN FARMING; Farming and Farming Economy, Historical and Practical. 3s.
142. MODERN FARMING: Stock—Cattle, Sheep, and Horses. 2s. 6d.
145. MODERN FARMING: Management of the Dairy—Pigs—Poultry. With Notes on the Diseases of Stock. 2s.
146. MODERN FARMING: Utilisation of Town Sewage—Irrigation—Reclamation of Waste Land. 2s. 6d.
   Nos. 140, 141, 142, 145, and 146 bound in 2 vols., cloth boards, 14s.

ARCHITECTURE AND BUILDING.

16. ARCHITECTURE, Orders of, by W. H. Leeds. 1s.

17. —————————— Styles of, by T. Talbot Bury. 1s. 6d.
 N.B. The Orders and Styles of Architecture in 1 vol., price 2s. 6d.

18. —————————— Principles of Design, by E. L. Garbett. 2s.

22. BUILDING, the Art of, by E. Dobson. 1s. 6d.

23. BRICK AND TILE MAKING, by E. Dobson. 2s.

25. MASONRY AND STONE-CUTTING, by E. Dobson. 2s. 6d.

30. DRAINAGE AND SEWAGE OF TOWNS AND BUILD-
 INGS, by G. D. Dempsey. 2s.
 (With No. 29, DRAINAGE OF LAND, 2 vols. in 1, 3s.)

35. BLASTING AND QUARRYING OF STONE, AND BLOW-
 ING UP OF BRIDGES, by Field-Marshal Sir J. F. Bur-
 goyne. 1s. 6d.

36. DICTIONARY OF TECHNICAL TERMS used by Architects,
 Builders, Engineers, Surveyors, &c. 4s.
 In cloth boards, 5s.; half morocco, 6s.

42. COTTAGE BUILDING, by C. B. Allen. 1s.

44. FOUNDATIONS AND CONCRETE WORKS, by E. Dobson.
 1s. 6d.

45. LIMES, CEMENTS, MORTARS, CONCRETE, MASTICS,
 &c., by G. R. Burnell. 1s. 6d.

57. WARMING AND VENTILATION, by C. Tomlinson. 3s.

83**. DOOR LOCKS AND IRON SAFES, by C. Tomlinson and
 Robert Mallet, C.E., F.R.S. 2s. 6d.

111. ARCHES, PIERS, AND BUTTRESSES, by W. Bland. 1s. 6d.

116. ACOUSTICS OF PUBLIC BUILDINGS, by T. R. Smith.
 1s. 6d.

123. CARPENTRY AND JOINERY, founded on Robison and
 Tredgold. 1s. 6d.

123*. —————————— ILLUSTRATIVE PLATES to the preceding.
 4to. 4s. 6d.

124. ROOFS FOR PUBLIC AND PRIVATE BUILDINGS,
 founded on Robison, Price, and Tredgold. 1s. 6d.

124*. IRON ROOFS of Recent Construction—Descriptive Plates. 4to.

127. ARCHITECTURAL MODELLING IN PAPER, Practical
 Instructions, by T. A. Richardson, Architect. 1s. 6d.

LONDON VIRTUE & CO., 26, IVY LANE.

128. VITRUVIUS'S ARCHITECTURE, translated by J. Gwilt, with Plates. 5s.

130. GRECIAN ARCHITECTURE, Principles of Beauty in, by the Earl of Aberdeen. 1s.

132. ERECTION OF DWELLING-HOUSES, with Specifications, Quantities of Materials, &c., by S. H. Brooks, 27 Plates. 2s. 6d.

156. QUANTITIES AND MEASUREMENTS; How to Calculate and Take them in Bricklayers', Masons', Plasterers', Plumbers', Painters', Paper-hangers', Gilders', Smiths', Carpenters', and Joiners' Work. With Rules for Abstracting, &c. By A. C. Beaton. 1s.

ARITHMETIC AND MATHEMATICS.

32. MATHEMATICAL INSTRUMENTS, THEIR CONSTRUC-TION, USE, &c., by J. F. Heather. 1s. 6d.

60. LAND AND ENGINEERING SURVEYING, by T. Baker. 2s.

61*. READY RECKONER for the Admeasurement of Land, Tables of Work at from 2s. 6d. to 20s. per acre, and valuation of Land from £1 to £1,000 per acre, by A. Arman. 1s. 6d.

76. GEOMETRY, DESCRIPTIVE, with a Theory of Shadows and Perspective, and a Description of the Principles and Practice of Isometrical Projection, by J. F. Heather. 2s.

83. COMMERCIAL BOOK-KEEPING, by James Haddon. 1s.

84. ARITHMETIC, with numerous Examples, by J. R. Young. 1s. 6d.

84*. KEY TO THE ABOVE, by J. R. Young. 1s. 6d.

85. EQUATIONAL ARITHMETIC: including Tables for the Calculation of Simple Interest, with Logarithms for Compound Interest, and Annuities, by W. Hipsley. In Two Parts, price 1s. each.

86. ALGEBRA, by J. Haddon. 2s.

86*. KEY AND COMPANION TO THE ABOVE, by J. R. Young. 1s. 6d.

88. THE ELEMENTS OF EUCLID, with Additional Propositions, and Essay on Logic, by H. Law. 2s.

90. ANALYTICAL GEOMETRY AND CONIC SECTIONS, by J. Hann. 1s.

91. PLANE TRIGONOMETRY, by J. Hann. 1s.

92. SPHERICAL TRIGONOMETRY, by J. Hann. 1s.
Nos. 91 and 92 in 1 vol., price 2s.

LONDON: VIRTUE & CO., 26, IVY LANE.

CIVIL ENGINEERING.

47. LIGHTHOUSES, their Construction & Illumination, by Alan Stevenson. 3s.

62. RAILWAY CONSTRUCTION, by Sir M. Stephenson. With Additions by E. Nugent, C.E.

62*. RAILWAY CAPITAL AND DIVIDENDS, with Statistics of Working, by E. D. Chattaway. 1s.

78. STEAM AND LOCOMOTION, on the Principle of connecting Science with Practice, by J. Sewell. 2s.

80*. EMBANKING LANDS FROM THE SEA, by J. Wiggins. 2s.

82**. A TREATISE ON GAS WORKS, AND THE PRACTICE OF MANUFACTURING AND DISTRIBUTING COAL GAS, by S. Hughes, C.E. 3s.

82***. WATER-WORKS FOR THE SUPPLY OF CITIES AND TOWNS, by S. Hughes, C.E. 3s.

118. CIVIL ENGINEERING OF NORTH AMERICA, by D. Stevenson. 3s.

120. HYDRAULIC ENGINEERING, by G. R. Burnell. 3s.

121. RIVERS AND TORRENTS, with the Method of Regulating their COURSE AND CHANNELS, NAVIGABLE CANALS, &c., from the Italian of Paul Frisi. 2s. 6d.

EMIGRATION.

154. GENERAL HINTS TO EMIGRANTS. 2s.
157. EMIGRANT'S GUIDE TO NATAL, by R. J. Mann, M.D. 2s.
159. EMIGRANT'S GUIDE TO NEW SOUTH WALES, WESTERN AUSTRALIA, SOUTH AUSTRALIA, VICTORIA, AND QUEENSLAND, by James Baird, B.A.
160. EMIGRANT'S GUIDE TO TASMANIA AND NEW ZEALAND, by James Baird, B.A.

FINE ARTS.

20. PERSPECTIVE, by George Pyne. 2s.
27. PAINTING; or, A GRAMMAR OF COLOURING, by G. Field. 2s.
40. GLASS STAINING, by Dr. M. A. Gessert, with an Appendix on the Art of Enamel Painting, &c. 1s.

LONDON: VIRTUE & CO., 26, IVY LANE.

41. PAINTING ON GLASS, from the German of Fromberg. 1s.
69. MUSIC, Treatise on, by C. C. Spencer. 2s.
71. THE ART OF PLAYING THE PIANOFORTE, by C. C. Spencer. 1s.

LEGAL TREATISES.

50. LAW OF CONTRACTS FOR WORKS AND SERVICES, by David Gibbons. 1s. 6d.
108. METROPOLIS LOCAL MANAGEMENT ACTS.   1s. 6d.
108*. METROPOLIS LOCAL MANAGEMENT AMENDMENT ACT, 1862; with Notes and Index. 1s.
109. NUISANCES REMOVAL AND DISEASES PREVENTION AMENDMENT ACT. 1s.
110. RECENT LEGISLATIVE ACTS applying to Contractors, Merchants, and Tradesmen. 1s.
151. A HANDY BOOK ON THE LAW OF FRIENDLY, INDUSTRIAL AND PROVIDENT, BUILDING AND LOAN SOCIETIES, by N. White. 1s.

MECHANICS & MECHANICAL ENGINEERING.

6. MECHANICS, by Charles Tomlinson. 1s. 6d.
12. PNEUMATICS, by Charles Tomlinson. New Edition. 1s. 6d.
33. CRANES AND MACHINERY FOR RAISING HEAVY BODIES, the Art of Constructing, by J. Glynn. 1s.
34. STEAM ENGINE, by Dr. Lardner. 1s.
59. STEAM BOILERS, their Construction and Management, by R. Armstrong. With Additions by R. Mallet. 1s. 6d.
63. AGRICULTURAL ENGINEERING, BUILDINGS, MOTIVE POWERS, FIELD MACHINES, MACHINERY AND IMPLEMENTS, by G. H. Andrews, C.E. 3s.
67. CLOCKS, WATCHES, AND BELLS, by E. B. Denison. New Edition, with Appendix to the 4th and 5th Editions. 3s. 6d.
    N.B.—Appendix (to the 4th and 5th Editions) sold separately, price 1s.
77*. ECONOMY OF FUEL, by T. S. Prideaux. 1s. 6d.
78*. THE LOCOMOTIVE ENGINE, by G. D. Dempsey. 1s. 6d.
79*. ILLUSTRATIONS TO THE ABOVE. 4to. 4s. 6d.
80. MARINE ENGINES, AND STEAM VESSELS, AND THE SCREW, by R. Murray. With Additions by E. Nugent.

82. WATER POWER, as applied to Mills, &c., by J. Glynn. 2s.

97. STATICS AND DYNAMICS, by T. Baker. 1s.

98. MECHANISM AND MACHINE TOOLS, by T. Baker; and TOOLS AND MACHINERY, by J. Nasmyth. With 220 Woodcuts. 2s. 6d.

113*. MEMOIR ON SWORDS, by Col. Marey, translated by Lieut.-Col. H. H. Maxwell. 1s.

114. MACHINERY, Construction and Working, by C. D. Abel. 1s. 6d.

115. PLATES TO THE ABOVE. 4to. 7s. 6d.

125. COMBUSTION OF COAL, AND THE PREVENTION OF SMOKE, by C. Wye Williams, M.I.C.E. 3s.

139. STEAM ENGINE, Mathematical Theory of, by T. Baker. 1s.

155. ENGINEER'S GUIDE TO THE ROYAL AND MER-CANTILE NAVIES, by a Practical Engineer. Revised by D. F. McCarthy. 3s

NAVIGATION AND SHIP-BUILDING.

51. NAVAL ARCHITECTURE, by J. Peake. 3s.

53*. SHIPS FOR OCEAN AND RIVER SERVICE, Construction of, by Captain H. A. Sommerfeldt. 1s.

53**. ATLAS OF 15 PLATES TO THE ABOVE, Drawn for Practice. 4to. 7s. 6d.

54. MASTING, MAST-MAKING, and RIGGING OF SHIPS, by R. Kipping. 1s. 6d.

54*. IRON SHIP-BUILDING, by J. Grantham. 2s. 6d.

54**. ATLAS OF 24 PLATES to the preceding. 4to. 22s. 6d.

55. NAVIGATION; the Sailor's Sea Book: How to Keep the Log and Work it off, &c.; Law of Storms, and Explanation of Terms, by J. Greenwood. 2s.

83 bis. SHIPS AND BOATS, Form of, by W. Bland. 1s. 6d.

99. NAUTICAL ASTRONOMY AND NAVIGATION, by J. R. Young. 2s.

100*. NAVIGATION TABLES, for Use with the above. 1s. 6d.

106. SHIPS' ANCHORS for all SERVICES, by G. Cotsell. 1s. 6d.

149. SAILS AND SAIL-MAKING, by R. Kipping, N.A. 2s. 6d.

LONDON: VIRTUE & CO., 26, IVY LANE.

PHYSICAL AND CHEMICAL SCIENCE.

1. CHEMISTRY, by Prof. Fownes. With Appendix on Agricultural Chemistry. 1s.

2. NATURAL PHILOSOPHY, by Charles Tomlinson. 1s.

3. GEOLOGY, by Major-Gen. Portlock. 1s. 6d.

4. MINERALOGY, by A. Ramsay, Jun. 3s.

7. ELECTRICITY, by Sir W. S. Harris. 1s. 6d.

7*. GALVANISM, ANIMAL AND VOLTAIC ELECTRICITY, by Sir W. S. Harris. 1s. 6d.

8. MAGNETISM, by Sir W. S. Harris. 3s. 6d.

72. RECENT AND FOSSIL SHELLS (A Manual of the Mollusca), by S. P. Woodward. 5s. 6d.

   N.B.—An Appendix by Ralph Tate, F.G.S., in the press.

79**. PHOTOGRAPHY, The Stereoscope, &c, from the French of D. Van Monckhoven, by W. H. Thornthwaite. 1s. 6d.

133. METALLURGY OF COPPER, by Dr. R. H. Lamborn. 2s.

134. METALLURGY OF SILVER AND LEAD, by Dr. R. H. Lamborn. 2s.

135. ELECTRO-METALLURGY, by A. Watt. 1s. 6d.

138. HANDBOOK OF THE TELEGRAPH, by R. Bond. 1s.

143. EXPERIMENTAL ESSAYS—On the Motion of Camphor and Modern Theory of Dew, by C. Tomlinson. 1s.
    THE HISTORY OF THE ELECTRIC TELEGRAPH, by Robert Sabine, F.S.A.

MISCELLANEOUS TREATISES.

112. DOMESTIC MEDICINE, by Dr. Ralph Gooding. 2s.

112*. A GUIDE TO HEALTH, by James Baird. 1s.

113. USE OF FIELD ARTILLERY ON SERVICE, by Taubert, translated by Lieut.-Col. H. H. Maxwell. 1s. 6d.

150. LOGIC, PURE AND APPLIED, by S. H. Emmens. 1s. 6d.

152. PRACTICAL HINTS FOR INVESTING MONEY: with an Explanation of the Mode of Transacting Business on the Stock Exchange, by Francis Playford, Sworn Broker. 1s.

153. LOCKE ON THE CONDUCT OF THE HUMAN UNDERSTANDING, Selections from, by S. H. Emmens. 2s.

LONDON: VIRTUE & CO., 26, IVY LANE.

NEW SERIES OF EDUCATIONAL WORKS.

[This Series is kept in three styles of binding—the prices of each are given in columns at the end of the lines.]

	Limp.	Cloth Boards.	Half Morocco.
	s. d.	*s. d.*	*s. d.*
1. ENGLAND, History of, by W. D. Hamilton	4 0	5 0	5 6
5. GREECE, History of, by W. D. Hamilton and E. Levien, M.A. . . .	2 6	3 6	4 0
7. ROME, History of, by E. Levien, M.A. .	2 6	3 6	4 0
9. CHRONOLOGY OF HISTORY, LITE-rature, Art, and Progress, from the earliest period to the present time . .	2 6	3 6	4 0
11. ENGLISH GRAMMAR, by Hyde Clarke, D.C.L.	1 0		
11*. HANDBOOK OF COMPARATIVE PHI-lology, by Hyde Clarke, D.C.L. .	1 0		
12. ENGLISH DICTIONARY, above 100,000 words, or 50,000 more than in any existing work. By Hyde Clarke, D.C.L. .	3 6	4 6	5 0
———————————, with Grammar		5 6	6 0
14. GREEK GRAMMAR, by H. C. Hamilton	1 0		
15. ——— DICTIONARY, by H. R. Hamilton. Vol. 1. Greek—English .	2 0		
17. ———————— Vol. 2. English — Greek	2 0		
——— Complete in 1 vol. . . .	4 0	5 0	5 6
———————————, with Grammar		6 0	6 6
19. LATIN GRAMMAR, by T. Goodwin, M.A.	1 0		
20. ——— DICTIONARY, by T. Goodwin, M.A. Vol. I. Latin—English .	2 0		
22. ———————— Vol. 2. English—Latin	1 6		
——— Complete in 1 vol. . . .	3 6	4 6	5 0
———————————, with Grammar		5 6	6 0
24. FRENCH GRAMMAR, by G. L. Strauss .	1 0		

LONDON: VIRTUE & CO., 26, IVY LANE.

	Limp.	Cloth Boards.	Half Morocco.
	s. d.	s. d.	s. d.
25. FRENCH DICTIONARY, by A. Elwes.			
Vol. 1. French—English	1 0		
26. ———— Vol. 2. English—French	1 6		
———— Complete in 1 vol.	2 6	3 6	4 0
————, with Grammar		4 6	5 0
27. ITALIAN GRAMMAR, by A. Elwes .	1 0		
28. ———— TRIGLOT DICTIONARY, by A. Elwes. Vol. 1. Italian—English—French	2 0		
30. ———— Vol. 2. English—French—Italian	2 0		
32. ———— Vol. 3. French—Italian—English	2 0		
———— Complete in 1 vol.		7 6	8 6
————, with Grammar		8 6	9 6
34. SPANISH GRAMMAR, by A. Elwes .	1 0		
35. ———— ENGLISH AND ENGLISH-SPANISH DICTIONARY, by A. Elwes	4 0	5 0	5 6
————, with Grammar		6 0	6 6
39. GERMAN GRAMMAR, by G. L. Strauss .	1 0		
40. ———— READER, from best Authors .	1 0		
41. ———— TRIGLOT DICTIONARY, by N. E. S. A. Hamilton. Vol. 1. English—German—French .	1 0		
42. ———— Vol. 2. German—French—English	1 0		
43. ———— Vol. 3. French—German—English	1 0		
———— Complete in 1 vol.	3 0	4 0	4 6
————, with Grammar		5 0	5 6
44. HEBREW DICTIONARY, by Dr. Bresslau. Vol. 1. Hebrew—English	6 0		
————, with Grammar	7 0		
46. ———— Vol. 2. English—Hebrew	3 0		
———— Complete, with Grammar, in 2 vols.		12 0	14 0
46*. ———— GRAMMAR, by Dr. Bresslau .	1 0		
47. FRENCH AND ENGLISH PHRASE BOOK	1 0		
48. COMPOSITION AND PUNCTUATION, by J. Brenan .	1 0		
49. DERIVATIVE SPELLING BOOK, by J. Rowbotham .	1 6		
50. DATES AND EVENTS. A Tabular View of English History, with Tabular Geography, by Edgar H. Rand.			

LONDON: VIRTUE & CO., 26, IVY LANE

GREEK AND LATIN CLASSICS,

With Explanatory Notes in English.

LATIN SERIES.

1. A NEW LATIN DELECTUS, with Vocabularies and Notes, by H. Young . . . . 1s.

2. CÆSAR. De Bello Gallico; Notes by H. Young . . 2s.

3. CORNELIUS NEPOS; Notes by H. Young . . . 1s.

4. VIRGIL. The Georgics, Bucolics, and Doubtful Poems; Notes by W. Rushton, M.A., and H. Young . 1s. 6d.

5. VIRGIL. Æneid; Notes by H. Young . . . . 2s.

6. HORACE. Odes, Epodes, and Carmen Seculare, by H. Young 1s.

7. HORACE. Satires and Epistles, by W. B. Smith, M.A. 1s. 6d.

8. SALLUST. Catiline and Jugurthine War; Notes by W. M. Donne, B.A. . . . . . . 1s. 6d.

9. TERENCE. Andria and Heautontimorumenos; Notes by the Rev. J. Davies, M.A. . . . . . 1s. 6d.

10. TERENCE. Adelphi, Hecyra, and Phormio; Notes by the Rev. J. Davies, M.A. . . . . . . 2s.

14. CICERO. De Amicitia, de Senectute, and Brutus; Notes by the Rev. W. B. Smith, M.A. . . . . 2s.

16. LIVY. Books i., ii., by H. Young . . . . 1s. 6d.

16*. LIVY. Books iii., iv., v., by H. Young . . . 1s. 6d.

17. LIVY. Books xxi., xxii., by W. B. Smith, M.A. . 1s. 6d.

19. CATULLUS, TIBULLUS, OVID, and PROPERTIUS, Selections from, by W. Bodham Donne . . . . 2s.

20. SUETONIUS and the later Latin Writers, Selections from, by W. Bodham Donne . . . . . . 2s.

21. THE SATIRES OF JUVENAL, by T. H. S. Escott, B.A., of Balliol College, Oxford.

LONDON: VIRTUE & CO., 26, IVY LANE.

GREEK SERIES.

1. A NEW GREEK DELECTUS, by H. Young . . 1s.

2. XENOPHON. Anabasis, i. ii. iii., by H. Young . . 1s.

3. XENOPHON. Anabasis, iv. v. vi. vii., by H. Young . 1s.

4. LUCIAN. Select Dialogues, by H. Young . . . 1s.

5. HOMER. Iliad, i. to vi., by T. H. L. Leary, M.A. . 1s. 6d.

6. HOMER. Iliad, vii. to xii., by T. H. L. Leary, M.A. 1s. 6d.

7. HOMER. Iliad, xiii. to xviii., by T. H. L. Leary, M.A. 1s. 6d.

8. HOMER. Iliad, xix. to xxiv., by T. H. L. Leary, M.A. 1s. 6d.

9. HOMER. Odyssey, i. to vi., by T. H. L. Leary, M.A. 1s. 6d.

10. HOMER. Odyssey, vii. to xii., by T. H. L. Leary, M.A. 1s. 6d.

11. HOMER. Odyssey, xiii. to xviii., by T. H. L. Leary, M.A. 1s. 6d.

12. HOMER. Odyssey, xix. to xxiv.; and Hymns, by T. H. L. Leary, M.A. . . . . 2s.

13. PLATO. Apologia, Crito, and Phædo, by J. Davies, M.A. 2s.

14. HERODOTUS, Books i. ii., by T. H. L. Leary, M.A. 1s. 6d.

15. HERODOTUS, Books iii. iv., by T. H. L. Leary, M.A. 1s. 6d.

16. HERODOTUS, Books v. vi. vii., by T. H. L. Leary, M.A. 1s. 6d.

17. HERODOTUS, Books viii. ix., and Index, by T. H. L. Leary, M.A. . . . . 1s. 6d.

18. SOPHOCLES. Œdipus Tyrannus, by H. Young . . 1s.

20. SOPHOCLES. Antigone, by J. Milner, B.A. . . . 2s.

23. EURIPIDES. Hecuba and Medea, by W. B. Smith, M.A. 1s. 6d.

26. EURIPIDES. Alcestis, by J. Milner, B.A. . . . 1s.

30. ÆSCHYLUS. Prometheus Vinctus, by J. Davies, M.A. . 1s.

32. ÆSCHYLUS. Septem contra Thebas, by J. Davies, M.A. 1s.

40. ARISTOPHANES. Acharnenses, by C. S. D. Townshend, M.A. . . . . 1s. 6d.

41. THUCYDIDES, Book i., by H. Young . . . . 1s.

LONDON: VIRTUE & CO., 26, IVY LANE.

1s. 6
1s. 6
1. 1s. 6
. 1s. 6
1s. 6
1s. 6
. 1s. 6
L.
. 2
4. 2
1s. 6d.
1s. 6d.
1s. 6d.

1s. 6d.
. 1s
. 2s
s. 6d.
1s.
1s.
1s.

6d.
1s.

VIRTUE & CO., 26, IVY LANE, PATERNOSTER ROW.

Euclid, the Elements of, with many additional Propositions, and Explanatory Notes; to which is prefixed an Introductory Essay on Logic. By HENRY LAW, C.E. 2s.

Analytical Geometry and Conic Sections, a Rudimentary Treatise on. By JAMES HANN, Mathematical Master of King's College School, London. 1s.

Plane Trigonometry, the Elements of. By JAMES HANN. 1s.

Spherical Trigonometry, the Elements of. By JAMES HANN. Revised by CHARLES H. DOWLING, C.E. 1s.

N.B.—The Elements of Plane and Spherical Trigonometry in one volume, price 2s.

Mensuration and Measuring, for Students and Practical Use. With the Mensuration and Levelling of Land for the purposes of Modern Engineering. By T. BAKER, C.E. New Edition, with Corrections and Additions by E. NUGENT, C.E. 1s. 6d.

Logarithms, a Rudimentary Treatise on: Mathematical Tables for facilitating Astronomical, Nautical, Trigonometrical, and Logarithmic Calculations; Tables of Natural Sines and Tangents and Natural Cosines. By HENRY LAW, C.E. 2s. 6d.

Rudimentary Astronomy. By the Rev. ROBERT MAIN, M.A., F.R.A.S., of the Royal Observatory, Greenwich. Illustrated. 1s.

Differential Calculus, the Elements of the. By W. S. B. WOOLHOUSE, F.R.A.S., F.S.S., &c. 1s.

The Measures, Weights, and Moneys of all Nations, and an Analysis of the Christian, Hebrew, and Mahometan Calendars. By W. S. B. WOOLHOUSE. 1s. 6d.

VIRTUE & CO., 26, IVY LANE, PATERNOSTER ROW.

CPSIA information can be obtained
at www.ICGtesting.com
Printed in the USA
LVHW081425190921
698212LV00010B/430

9 780342 201662